CDM 2015

A Practical Guide for
Architects and Designers

CDM 2015 – A Practical Guide for Architects and Designers

© Scott Brownrigg, 2015

Published by RIBA Publishing, part of RIBA Enterprises Ltd,
The Old Post Office, St Nicholas Street, Newcastle upon Tyne, NE1 1RH

ISBN 978 1 85946 613 1

Stock code 84369

The rights of Scott Brownrigg to be identified as the Authors of this Work have been asserted in accordance with the Copyright, Designs and Patents Act 1988 sections 77 and 78.

All rights reserved. No part of this publication may be reproduced, stored in a retrieval system, or transmitted, in any form or by any means, electronic, mechanical, photocopying, recording or otherwise, without prior permission of the copyright owner.

British Library Cataloguing-in-Publication Data
A catalogue record for this book is available from the British Library.

Publisher: Steven Cross
Commissioning Editor: Sarah Busby
Project Editor: Richard Blackburn
Cover Design: Tom Cabot/Ketchup
Designed by Gavin Ambrose
Typeset by Academic+Technical, Bristol, UK
Printed and bound by Page Bros, Norwich, UK

While every effort has been made to check the accuracy and quality of the information given in this publication, neither the Author nor the Publisher accept any responsibility for the subsequent use of this information, for any errors or omissions that it may contain, or for any misunderstandings arising from it.

RIBA Publishing is part of RIBA Enterprises Ltd.
www.ribaenterprises.com

CDM 2015
A PRACTICAL GUIDE FOR ARCHITECTS AND DESIGNERS

Paul Bussey

RIBA Publishing

This book is dedicated to my mother, Joan Iris Bussey, who died unexpectedly during its development.

Foreword

Health and safety in the design world is challenging. Making the right decisions, in proportion to all the other issues that have to earn their rightful place, taxes even the most skilled. If this were not enough of a mountain to climb we have an industry that is highly fragmented with many, sometimes opposing, viewpoints.

I have known Paul for many years and our approach has always been to identify the appropriate path for the designer. Ours often feels a lone voice. One that, unless you are a designer, can be misunderstood. However, in recent years several significant experts across the construction industry have agreed that there is a need for designers to look at this in a way only they can; to ignore the noise; to challenge convention; to innovate; and to be the instigators of delivery. This achieves real benefits on the ground. This view has been widely supported by the HSE and together with the emergence of CDM 2015 makes this book very timely indeed.

Many were greatly influenced by Andrew Townsend and his unique take on this area. His untimely death has underlined the need to carry his baton and in many ways this book is exactly that. Building on the techniques Paul has developed and tested in practice it is a testament to approaching the subject in a practical, balanced and proportionate way.

This book is a must for all architects and designers practising in the built environment today. It has ideas and guidance learned in the face of commercial and legal realities.

I congratulate Paul and hope this will encourage many more to understand that designers need to take a leadership role, look at health and safety differently, own the space, and provide those constructing, using and maintaining buildings with the best and safest solutions possible.

Peter L. Caplehorn
Deputy Chief Executive and Policy Director
Construction Products Association

Acknowledgements

The author would like to thank the following for their help with this book.

Southampton Solent University
Interserve Building
Scott Brownrigg Board, Staff and Directors
Tony Putsman
Andrew Townsend (in memorium)
Sarah Busby
Peter Caplehorn
North Hertfordshire College
Helen Taylor
DIOHAS Members
CIC Expert Health and Safety Panel
Various HSE Committees and Working Groups
RIBA Regulations and Standards Group
The Wren Insurance Technical Forum
Andy Jobling
Association for Project Safety

About the author

Paul Bussey is a Chartered Architect and Associate at Scott Brownrigg and Design Delivery Unit, a Scott Brownrigg Company. He has some 35 years of experience as an architect, for the last 20 of which he has worked with architects and the wider industry to apply the Construction (Design and Management) Regulations to design projects in a meaningful and creative way. He writes widely on CDM for the Association for Project Safety (APS), the Royal Institute of Chartered Surveyors (RICS) and various learned journals, and provides continuing professional development for architects on implementing CDM for the Royal Institute of British Architects (RIBA). He is also a visiting lecturer at various universities. He has worked closely with the industry's leading professional bodies over the last ten years on the development of CDM guidance and best practice, including the Health and Safety Executive (HSE), the APS, Safety In Design, British Standards Institution (BSI), the Construction Industry Council (CIC), Design Best Practice, CIRIA and the Department for Communities and Local Government (DCLG). He is also the health and safety expert on the RIBA Regulations and Standards and CPD Groups. In 2014 he convened the Healthy Design and Creative Safety Symposium at Sheffield University, and is particularly interested in developing easy-to-use and visual tools to support designers and design teams to achieve design excellence while excelling in health and safety.

About Design Delivery Unit, a Scott Brownrigg Company

Design Delivery Unit is an architectural delivery company owned by an international design practice, Scott Brownrigg. It provides a range of specialist services to clients including the role of executive architect, delivering projects from RIBA Stage 3, post-planning, working with contractors and developers, and in partnership with concept architects. Other services include design optimisation to add value to and safeguard the concept architect's design intent; auditing and development of a project to ensure compliance with design standards and planning obligations; leading the consultant team to develop the technical design and deliver production information suitable to build from; and participation in the procurement and construction process, with particular expertise in design and build construction management and traditional contracts.

As well as being a leading building information modelling (BIM) adopter, using Revit Architecture to deliver all its projects, Design Delivery Unit provides additional support services including CDM consultancy, a BIM coordinator role, interior design and professional due diligence.

This support is backed up with the latest industry expertise and insight provided by experienced technical staff with industry affiliations, such as: RIBA Council, RIBA Practice and Profession Committee, APS, BSI, CIC, CONIAC, DCLG, DIOHAS, HSE, RIBA CIAT BRE Technical Taskforce, and BRE Global Panel of Experts.

As this book is intended to help to save lives and to prevent injuries through good design, we (the author and Scott Brownrigg) have decided to pledge 50% of the royalties to Article 25, a humanitarian architectural charity.

Contents

1.	Introduction: A practical and creative approach to the integration of CDM 2015 into architectural design	2
2.	The EU Directive and the need for the CDM Regulations	8
3.	The UK regulatory CDM framework within which we work and the reasons for change	16
4.	CDM 2007 and CDM 2015 compared and contrasted	34
5.	Positive behaviours, cultures and approaches	74
6.	Hazard awareness and risk identification	98
7.	Conclusions: CDM Differently	142
Appendix I	Professional competence, support and information	145
Appendix II	Healthy Design and Creative Safety – methods of teaching and learning about CDM	149
Appendix III	Pre-2007 research by the HSE into construction accidents	152
References and bibliography		161
Index		165
Image credits		168

1 Introduction: A practical and creative approach to the integration of CDM 2015 into architectural design

A practical and creative approach

This guide introduces and explains the new CDM Regulations 2015. It outlines the key changes, who the duty holders are and – in clear tables allowing easy comparison – how the responsibilities of each duty holder have changed from 2007 to 2015.

However it also introduces a practical and creative approach to the integration of health and safety or construction design and management (CDM) into architectural projects in a proportionate and practicable way. It is intended as a regular reference source and introduces the reader to a number of other research documents, legislation and information sources with which they need to be familiar to deliver their new roles under the CDM Regulations. The need for each role is explained, along with the basic knowledge required. The intention is to promote creative integration of CDM into architectural design in a manner that is artistically and intellectually stimulating and collaborative, while meeting the legislative and ethical objectives of CDM.

This document is primarily intended for individual architects or aesthetic designers in large or small practices, as well as academics and students in the architectural and construction sectors. Designers from civil, structural and mechanical engineering or surveying backgrounds and other interested groups such as clients, contractors and legislators may find the guide helpful in understanding the rationale, motivations and methods of the architectural design community. It should therefore enable a range of construction professionals and designers to work together in an environment of mutual respect and design collaboration for better and safer architecture.

A moral and ethical approach

The guide also promotes a moral and ethical approach to integrating health and safety into everyday design activities, simply and proportionately, without fear of criminal or civil recrimination. This needs to happen while also fulfilling the concept design and brief requirements, and in parallel with normal day-to-day business activities.

Designers cannot entirely prevent all accidents or instances of ill-health, but they can make a contribution to the prevention process and learn how to better protect construction and maintenance operatives, building users and others. Unfortunately there are many myths and misconceptions about this process that have developed since the introduction of CDM in 1994. This book will help to counter these misunderstandings and allow designers and their teams to provide 'proportionate and practicable' solutions, processes and information.

The moral and ethical requirement for all professionals, and society in general, to consider health and safety in all our actions is at the heart of the integration of CDM into architectural design and construction.

The impact of our actions on others should be considered in every exchange between project team members, particularly in terms of fair-mindedness, integrity, honesty, self-knowledge and also an understanding that our behaviour has consequences for the welfare of others. This approach is fundamental to the better integration of CDM into architectural design and construction. The processes outlined here should help to prevent the dysfunctional and risk-averse attitudes of project teams of the last 20 years, and encourage a collaborative team environment and approach in the future.

Integration of health and safety

For most of us, the integration of health and safety into our daily lives is an ever present but subconscious process. Achieving a balance between safety and activity is a continual source of debate, amusement, misunderstanding and, quite often, disagreement. Carrying this theme across to a construction project, there will often be team members with little or no aesthetic training, who struggle to understand the visual, cultural and artistic priorities of designers. Meanwhile designers can be guilty of assuming that all construction and maintenance-related decisions are solely the responsibility of the contractor or end user. This situation has led led to many failings in construction projects and completed buildings across the industry. These polarised positions on safety issues can be irreconcilable and may cause considerable disruption to design team agreements and productivity. This disconnect is at the very heart of the need for this 'practical and creative guide' which will explore the fundamental relationship between construction safety and design, and propose methods for successfully integrating them while creating the opportunity for excellence both in terms of architecture and safety.

The UK architectural profession and CDM

Perhaps due to the relatively recent introduction of the CDM Regulations teaching of the implications of health and safety on design, although part of the RIBA and ARB curriculum, could be far better integrated into architectural education. For this reason it is not yet as well embedded as it might be in architectural practice, although annual CPD on the topic is mandatory for RIBA chartered members. This issue is being addressed by the RIBA. See Appendix II.

There is a lot of other guidance that describes the regulatory landscape; meanwhile this guide is intended to put the Regulations into a creative context for designers, providing a working document to support teaching, learning or practising architecture. It is hoped that a growing language and methodology of design safety will develop that relates to our own design profession, but at the same time is acceptable to and intelligible by other construction, legal and health and safety professionals.

Demonstrating good practice

This book aims to bring the Regulations, research and methods of integrating safety into design to the attention of the entire design and construction community to empower the project team to make appropriate health and safety design decisions.

It explains the CDM Regulations 2015 in a clear and accessible way for architects and designers, and provides practical solutions and case studies as exemplars for the resolution of health and safety issues on projects. It is hoped that this can begin an industry-wide or profession-based framework for providing examples of good practice for mutual benefit. This is also intended to reconcile the sometimes divergent aspirations of the industry on health and safety by examining the facts and various research reports that might have led to misunderstandings (see Appendix III). The guide cannot answer all of the health and safety questions levelled at the architectural design community, but it will hopefully provide a start.

Information overload

There are an enormous number of British Standards, Approved Codes of Practice and guidance documents that provide very detailed research evidence and recommendations for the industry. However the sheer volume and cost of such information often leaves the designer and contractor either with information overload or a lack of suitable information. The accessibility of unmoderated information on the internet can also be a problem, as it cannot possibly all be effectively assimilated. As a response to this situation a number of global organisations have started to compile easily searchable databases of good practice guidance and case studies for their own use, and also to inform and assist the wider design and construction community. It is hoped that the professions will be able to capture this information through collation of concise peer-group contributions and web-accessible databases, available to all. This guide provides a route map to this developing spectrum of information.

Note: where information is written by the author or is a commentary on the primary source-referenced documents, the text and annotations are red.

2 The EU Directive and the need for the CDM Regulations

Chapter overview

- Understanding the CDM Regulations
- Why do we need to acknowledge the EU Directive?
- The CDM Regulations 1994 and 2007
- What was the HSE view of CDM in 2007?

Understanding the CDM Regulations

Why do we need the CDM Regulations?
While this book provides an essential summary of the context and development of the UK CDM Regulations, it is not intended to provide a detailed history, which can be found elsewhere (see References and bibliography). Twenty years have passed since the introduction of the Construction (Design and Management) Regulations (CDM) and there has just been a second revision to implement the requirements of the European Parliament and to better embed these moral and ethical requirements into the construction industry.

A key intention is to explain the essence of CDM 2015 in comparison to its predecessors.

Various actions can have serious unintended consequences, including injury and even death. The Regulations set out to support the coordinated workforce of client, designer and contractor to put together plans, select capable people to work together and produce systems of work to deliver successful architecture. Unfortunately this outcome is not universal, showing the need for renewed consideration of the Regulations.

Proposed changes in 2015
Among the barriers to improvements in the construction industry are the principles set up at the beginning

of projects, such as health and safety policies, pre-qualification questionnaires and competence schemes. These provide an illusion of safety but in fact burden designers, construction managers and site operatives with increased paperwork to demonstrate compliance, without necessarily bringing improvements. When an accident occurs a further paperchase is embarked on to apportion blame, and additional layers of paperwork added in an attempt to prevent it happening again. This is a self-perpetuating cycle which seems to have no purpose beyond litigation.

However, after a false start in 2007, the HSE has in 2015 finally attempted, with government support, to stop unnecessary bureaucracy and promote a 'proportionate approach' to design and construction projects. This is a welcome attempt to assist small and medium enterprises in particular to take appropriate steps to do their job: to deliver beautiful buildings, meet client requirements and earn a living while facilitating the safety of others.

A change in attitude
There is much discussion and concern in the industry surrounding these proposals. The HSE now accepts integration of the Regulations over a six-month transitional period to ease the process of change, but the aspiration to reduce paperwork, while minimising the risk of litigation and accidents in a proportionate manner, will require considerably more patience and sharing of good practice. CDM 2007 was full of terms open to interpretation such as 'proportionate approach to risk', 'suitable and sufficient [actions]' and 'so far as is reasonably practicable' and these terms appear again in CDM 2015. However there is now a greater move to define these 'relative' terms with more tangible examples and definitions. The often-expressed concern that 'we will only know if we took appropriate action when it is decided in court' is not conducive to inspiring change, so these definitions are vital to create a shared understanding and approach across the industry.

The whole industry must therefore work together to effect change. There may be resistance but we must try; it is the designer institutes of the architects and civil engineers who can set the standards and level of acceptable practice.

Why do we need to acknowledge the EU Directive?

The EU Directive underpins the original introduction of the CDM Regulations in 1994/5 and is still relevant today. In spite of its imminent review, the need to copy out this legislation in CDM 2015 is a primary aim of the government and HSE in order to avoid financial penalties. There is also a requirement to improve safety in design considerations. The Directive states:

> … that the Council shall adopt, by means of directives, minimum requirements for encouraging improvements, especially in the working environment, to ensure a better level of protection of the safety and health of workers [because]:
>
> … unsatisfactory architectural and/or organizational options or poor planning of the works at the project preparation [i.e. design] stage have played a role in more than half of the occupational accidents occurring on construction sites in the Community.[1]

While the industry has targeted designers on the basis of this report, this is actually a misunderstanding. Rather than blaming the architectural community, the Directive simply suggests that it has been partially responsible for accidents and that there should be greater definition of all the roles involved at the early design stages. It goes on to state that:

> … when a project is being carried out, a large number of occupational accidents may be caused by inadequate coordination [i.e. at the design and construction stages], particularly where various undertakings work simultaneously or in succession at the same temporary or mobile construction site … it is therefore necessary to improve coordination between the various parties concerned at the project preparation stage and also when the work is being carried out.[1]

Arguably construction-stage site coordination is now operating well on the larger sites but not on the smaller ones. However the coordination of design at all stages is perceived by many, including the HSE, as not working as intended. This was a key part of the original motivation for the introduction of the CDM Regulations in 1994, and the two later iterations in 2007 and now 2015.

The CDM Regulations 1994 and 2007

The 1994 Regulations originally set the scene for our CDM legislative landscape, but it is worth looking again at the holistic development of CDM and how it was received by the experts.

The HSE commissioned a number of research reports on CDM 1994 in the years running up to CDM 2007. They tried to pin down why there were still such a large number of accidents in the industry and how designers could improve their knowledge and skills to reduce them. This review of the 1994 Regulations was seen as a call to arms for CDM 2007, although some felt that it was overly critical of designers and their role. These reports are included in Appendix III with analysis for more detailed scrutiny by the institutions and experts with respect to their validity and need to be addressed.

The paper below, produced by the CONIAC Secretariat in 2009, evaluates the feedback on the 1994 Regulations that directly influenced CDM 2007.

Early feedback on CDM 94 suggested that it was failing to promote effective health and safety but was prompting wasteful bureaucracy and related burdens on business. Industry-wide consultation in 2002 disclosed support for the principles of CDM 94 but dissatisfaction with their implementation in the Regulations. Among the criticisms offered were that the Regulations were inflexible and difficult to understand, that the duties of clients should be more proportionate to their level of influence, and that the Planning Supervisor role was ineffective in many ways. Subsequently, the [former] Health and Safety Commission approved a project to revise the Regulations, and laid down the following objectives for it:

(a) Simplifying the Regulations to improve clarity – so making it easier for duty holders to know what is expected of them.
(b) Maximising their flexibility – to fit with the vast range of contractual arrangements in the industry.
(c) Making their focus planning and management, rather than the plan and other paperwork – to emphasise active management and minimise bureaucracy.
(d) Strengthening the requirements regarding coordination and cooperation, particularly between designers and contractors – to encourage more integration.
(e) Simplifying the assessment of competence (both for organisations and individuals) to help raise standards and reduce bureaucracy.[2]

The aspirations for CDM 1994 were certainly well intentioned, but for a variety of reasons the industry did not fully embrace these underlying intentions. Instead it created further bureaucratic procedures for others to carry out, or found work-arounds or shifted responsibility. Overall the number of construction workplace deaths did reduce, but there were still unacceptable levels of long-term health issues, often resulting in long-latency effects and deaths.

Consequently the HSE, in collaboration with the industry (most particularly with the assistance of a CONIAC working group), embarked on a revision of the Regulations. This was designed to address the identified deficiencies according to the government's Better Regulation principles, and thus CDM 2007 came into force on 6 April 2007. During a parliamentary debate on the Regulations following their introduction, ministers agreed that the HSE should carry out an early review of CDM 2007 (i.e. within three years.)

This extract from the introduction of the 2007 CDM Approved Code of Practice clearly states that its aims were linked to the evaluation of the 1994 Regulations:

The key aim of CDM 2007 is to integrate health and safety into the management of the project and to encourage everyone involved to work together to:

(a) improve the planning and management of projects from the very start;
(b) identify hazards early on, so they can be eliminated or reduced at the design or planning stage and the remaining risks can be properly managed;
(c) target effort where it can do the most good in terms of health and safety; and
(d) discourage unnecessary bureaucracy.

These Regulations are intended to focus attention on planning and management throughout construction projects, from design concept onwards. The aim is for health and safety considerations to be treated as an essential, but normal part of a project's development – not an afterthought or bolt-on extra. The effort devoted to planning and managing health and safety should be in proportion to the risks and complexity associated with the project. When deciding what you need to do to comply with these Regulations, your focus should always be on action necessary to reduce and manage risks. Any paperwork produced should help with communication and risk management. Paperwork which adds little to the management of risk is a waste of effort, and can be a dangerous distraction from the real business of risk reduction and management.[3]

Three years after implementation of CDM 2007 an evaluation and review of its success was undertaken, and this is documented in the next chapter.

What was the HSE view of CDM 2007?

The following speech was given by Judith Hackitt, Chair of the HSE, in 2010, at the NEBOSH graduation. It encapsulates the state of the CDM 2007 landscape and the HSE view on the latest regulations at that time:

> … it should be a statement of the obvious that every single person in the workplace should be able to go home at the end of every day unharmed by their work. But between 150 and 200 people continue to die in workplace incidents every year. More than 100,000 suffer over-three-day lost-time injuries – over 20,000 of these are really serious injuries which may result in people never working again … the number of people who die early from diseases caused or made worse by work is close to 10,000.
>
> But in spite of this, here in the UK we have a better track record than most. … We have arrived at this point by being good at managing workplace health and safety. We have a strong, non-prescriptive regulatory framework enacted by the Health and Safety at Work Act. Legislation that is as fit for purpose today as it's ever been. We have a proportionate approach which recognises that risks can be neither eliminated nor avoided but can be managed reasonably and sensibly.
>
> We have very many businesses who understand that proportionate approach. And who also understand that businesses who are good at health and safety are also good at managing and motivating their staff and are, as a result, successful in business terms.
>
> This world-leading position comes as a consequence, in part, of the work of many people like ourselves who have acquired expertise in occupational safety and health through study and who are able to apply that in-depth knowledge.
>
> I am talking about application in the spirit of recognising that good performance comes from:
>
> — proportionate application of knowledge to tackle the real risks;

> — using knowledge to inspire and guide others to do the right thing as simply and easily as possible, not by weighing them down with bureaucracy and paperwork; and
> — by getting others to share our enthusiasm for ensuring that people do go home from work every day unharmed, because it's morally right – not because of fear of breaking rules, or of prosecution or simply because health and safety zealots tell them to ... Apply your knowledge with care and sensitivity. Use it to inspire and motivate others, not to burden them with red tape.[4]

Such an impassioned statement has given the author and others within Designers' Initiative on Health and Safety (DIOHAS) and the RIBA added motivation to pursue similar aims.

3

The UK regulatory CDM framework within which we work and the reasons for change

Chapter overview

- The main changes introduced in CDM 2007
- An independent government report on health and safety in 2011
- The professions' response
- Understanding risk and its 'tolerability'
- 'General Principles of Prevention' and examples of their application for CDM 2015

The main changes introduced in CDM 2007

In order to understand the motives and intentions behind the new 2015 Regulations it is essential to understand the key issues introduced in CDM 2007. Below is a summary of the main changes taken from the CDM 2007 Approved Code of Practice, with comments from the author:

1. CDM 94 and the Construction (Health, Safety and Welfare) Regulations 1996 (CHSW 96) consolidated into a single set of Regulations.
2. The Regulations were grouped by duty holder, so it was easier for each to see what their duties were. Unfortunately this tends to discourage a collaborative team approach.
3. Projects for domestic clients no longer needed to be notified. But CDM still applied to design and construction issues on all projects, a fact that was misunderstood by this notification exemption.
4. The Regulations applied to all sites, but there were additional duties for sites where construction work lasted more than 30 days or took more than 500 person days. This was widely misunderstood as meaning that CDM did not apply to non-notifiable projects.
5. Provision for a 'client's agent' was removed. But many projects include very influential project managers or lead designers who need to be integrated with the CDM process.
6. The planning supervisor role ceased to exist. A CDM coordinator was introduced to advise and assist the client; to coordinate the planning and design phase and to prepare the health and safety

file. This tended to be a peripheral client advisor monitoring role rather than a facilitator of CDM integration into projects.
7. *There was a simple trigger for the appointments of the CDM coordinator and the principal contractor, and preparation of a written health and safety plan. This trigger was the same as the notification threshold, i.e. 30 days or 500 person days of construction work.* This caused confusion around applicability and additional duties.
8. *Demolition was treated in the same way as any other construction activity, except that a written plan was required for all demolition work.*
9. *Clearer guidance is given in the ACOP on competence assessment, which it is hoped will save time and reduce bureaucracy.* This was a major failure and a cause of increased bureaucracy in CDM 2007, and will hopefully be clarified and rectified in CDM 2015.
10. *There was an enhanced client duty (making explicit duties which already existed under the Health and Safety at Work Act 1974 and the Management of Health and Safety at Work Regulations 1999) to ensure that the arrangements other duty holders had made were sufficient to ensure the health and safety of those working on the project.* This is being clarified in CDM 2015.
11. *A new duty was placed on designers to ensure that any workplace that they designed complied with relevant sections of the Workplace (Health, Safety and Welfare) Regulations 1992.* This has over-complicated CDM despite the laudable intention of putting all construction health and safety considerations into one set of regulations.
12. *Clients and contractors (including the principal contractor) must tell those they appoint how much time they have allowed, before work starts on site, for appointees to plan and prepare for the construction work.*[5]

Due to the requirement for a three-year evaluation and review of CDM 2007 and other health and safety regulations, various reports were commissioned by the government and industry, and these are discussed below.

An independent government report on health and safety in 2011

As a result of conflicting industry views, and the generally poor perception of health and safety in wider society, the government commissioned an independent health and safety expert, Professor Löfstedt, to provide a wide-ranging report on CDM 2007. Professor Löfstedt stated the following in relation to CDM 2007:

> In general, the problem lies less with the regulations themselves and more with the way they are interpreted and applied. In some cases this is caused by inconsistent enforcement by regulators and in others by the influences of third parties that promote the generation of unnecessary paperwork and a focus on health and safety activities that go above and beyond the regulatory requirements. Sometimes the legislation itself can contribute to the confusion, through its overall structure, a lack of clarity, or apparent duplication in some areas.
>
> The 'so far as is reasonably practicable' qualification in much of health and safety legislation was overwhelmingly supported by those who responded to the call for evidence on the grounds that it allows risks to be managed in a proportionate manner.
>
> However, there is general confusion over what it means in practice and many small businesses find it difficult to interpret.
>
> Meanwhile, there are instances where regulations designed to address real risks are being extended to cover trivial ones, whilst the requirement to carry out a risk assessment has turned into a bureaucratic nightmare for some businesses. The legal requirement to carry out a risk assessment is an important part of a risk management process but instead businesses are producing or paying for lengthy documents covering every conceivable risk, sometimes at the expense of controlling the significant risks in their workplace.
>
> As he said 'the sheer mass of regulation is a key concern for many businesses. Although there is considerably less regulation than 35 years ago, businesses still feel that they have to work through too many regulations or use health and safety consultants.' He also suggested 're-directing enforcement activity towards businesses where there is the greatest risk of injury or ill health'.

His conclusions were that *'the general sweep of requirements set out in health and safety regulations are broadly fit for purpose but there are a few that offer little benefit to health and safety and which the Government should remove, revise or clarify...'*. Thus as well as the revision of CDM 2007 and the associated ACOP to ensure a *'clearer expression of duties, a reduction of bureaucracy and appropriate guidance for small projects'* he also suggested a review of the Work at Height Regulations 2005 to ensure they didn't go *'beyond what was proportionate'*. Overall this would ensure *'the regulatory and legal system are better targeted towards risk and support the proper management of health and safety instead of a focus on trying to cover every possible risk and accumulating paperwork.'*

He also felt there was a *'need to stimulate a debate about risk in society to ensure that everyone has a much better understanding of risk and its management'* which should include a House of Lords Select Committee and an expert group set up by the Chief Scientific Advisor to the government. In parallel to this he recommended a requirement for the HSE to *'help businesses understand what is "reasonably practicable" for specific activities where the evidence demonstrates that they need further advice to comply with the law in a proportionate way'*, although a strategy for this has not yet been outlined.[6]

Professor Löfstedt therefore recommended a programme of 'sector-specific consolidations', meaning simplification and streamlining of regulations, to be completed by April 2015.

His findings seemed to ring true, as the way regulations are applied by over-zealous practitioners is a common complaint among designers, along with unnecessary bureaucracy and confusion over the definition of significant risk versus trivial risk. It was clear that greater clarity around 'so far as is reasonably practicable' (SFARP) and proportionate risk management was required across the industry. It also recognised that information overload was rife, and caused more problems than it solved.

The professions' response

As a result of Professor Löfstedt's report the Construction Industry Council (CIC) and Institute of Civil Engineers (ICE), representing the design professions, formally responded as follows in 2010:

Recommendations from this CIC/ICE response:

- All regulations and ACOPs need to be regularly reviewed with input from professionals working in that field
- In places, Regulations and, more generally, ACOPs need to be simplified and/or their requirements clarified to reduce opportunities for misinterpretation – with input from professionals working in that field
- An urgent review is required to bring improved definition to 'reasonably practicable'
- A specific review is required to consider the application of 'reasonably practicable' as applied to construction designers

Reasonably practicable

The extension of this principle to construction designers under the CDM Regulations (Regulation 11) has brought uncertainty (often accompanied by significant bureaucracy) to the point where the working group believes the Regulation is discredited and un-enforceable. Generally speaking, Designers are remote from the manual tasks (of construction or maintenance), have no control over it, and no contact with those undertaking the work. This was not the scenario envisaged when the test was originally introduced. We support the concept, but the manner in which it has been introduced as a statutory obligation has shown itself to be unworkable. The legal definition of 'so far as is reasonably practicable' has been established; however, as shown by ICE's study it offers no assistance in a practical sense. There is no case law which offers any substantive clarification for construction Designers.

Therefore the key issue remains 'how far is far enough?'

After 16 years of its existence, no-one in industry knows how to apply 'Designers Duties' in a practical project related manner in order to satisfy the law. As mentioned, there is no substantive case law to assist this practical understanding. Whilst there are examples of good practice, that is not necessarily the same as compliance in the absence of endorsement *by the Health and Safety Executive. As a consequence of this position, a significant number of Designers ignore the requirement; a further significant number waste valuable time and energy producing unnecessary or worthless paperwork, i.e. the ubiquitous 'Risk Assessment', for no benefit. Those who wish to encourage compliance are prevented from doing so as they cannot be sure what 'compliance' looks like. At a time when industry should be looking for efficiency and added value this is an unacceptable and wasteful use of resource and a missed opportunity for what is essentially a good concept.*

This issue is not concerned with a peripheral requirement, or some issue of interest only to a few. It goes to the heart of the design process, affecting not only designers but also CDM Co-ordinators who are unable to be satisfied that designers are discharging their obligations. For such major players in such an important industry to have this gross uncertainty is unacceptable.

Recommendation:

'Designers Duties' and its accompanying ACOP text, needs to be re-drafted to allow clear explanation, execution and enforcement. Such action would have a significant beneficial effect upon our built infrastructure: in terms of reducing accidents and ill health, but also in terms of improved projects over their lifespan.[7]

This response reaffirms Professor Löfstedt's contention that the definition and understanding of SFARP is key to the proportionate and practicable implementation of CDM at design stages.

Definition of SFARP

Based on research and precedent from the Health and Safety at Work Act 1974, case law and the HSE document 'Reducing risks, protecting people 2001', and the view of other authoritative sources, the author's interpretation of SFARP for design stages is as follows:

'*In essence, it requires weighing the risk against the resources needed to eliminate or reduce the risk. The question of whether a measure is or is not "reasonably practicable" is one which requires no more than the making of a value judgement in the light of all the other influencing factors (not just H&S). This approach empowers the team to make judgements based on all the information available without fear of prosecution.*'

This interpretation should be the subject of further and rigorous discussion within the RIBA and other institutions and associations such as ICE, CIC and APS, to agree on the acceptability of this definition.

The CDM 2015 draft *Legal (L) Series guidance* provides a definition of 'reasonably practicable' in its glossary:

> This means balancing the level of risk against the measures needed to control the real risk in terms of money, time or trouble. However, you do not need to take action if it would be grossly disproportionate to the level of risk.[8]

Although this is encouraging, there is still considerable ambiguity with this definition and more discussion of what constitutes proportionate action for all duty holders and others is needed.

Understanding risk and its 'tolerability'

Discussion about 'reasonably practicable' and introducing the concept of 'tolerability of risk' had been instigated by an earlier 2001 HSE report, 'Reducing Risks, Protecting People' (R2P2), but was not clearly expressed in the CDM 2007 Regulations. It is thought that this was because the concept was considered too complex for the general industry. However, arguably, its exclusion has caused more confusion and misinterpretation than any other single factor. Timothy Walker, then Director General of the Health and Safety Executive, explained the concepts within this document in his speech in January 2003 at University College London on 'Decision Making on the Basis of Risk':

> In partnership with its stakeholders, the Health and Safety Executive (HSE) has the task of ensuring that risks to people's health and safety from work activities are properly controlled. However, resources are not unlimited; they need to be used where they will do most good. We seek control regimes, therefore, that are proportionate to the risk, not 'hammers to crack nuts'. In consequence, there are decisions to be made about the deployment of resources which involve the trading-off of the costs of control against the reduction in harm it achieves.
>
> **Reducing Risks, Protecting People** – widely known as R2P2. In this document HSE addresses its stakeholders and sets out how, in consultation

with them, it decides to regulate. R2P2 is an important document because it makes transparent the protocols and procedures we follow to ensure that the process of decision-making, including risk assessment and risk management, is perceived as valid.

R2P2 describes the tolerability of risk (TOR) framework (see **Figure 3.1**) which provides criteria for HSE to apply to its assessments of risk so as to reach a decision on regulatory control. The framework is constructed to reflect how people in general approach risk, that is, some risks are so high that they would be viewed as intolerable whatever the benefits that might be gained by taking the risk. Other risks are seen as too small to be of any further concern. In between are risks at a level where they are of concern but can be tolerated, provided that they are reduced to 'as low as reasonably practicable' (ALARP).

The ALARP suite of guidance – the HSE addresses its own staff about what they should expect to see in duty holders' demonstrations that risk has been reduced to 'as low as reasonably practicable' (ALARP). The test of 'reasonable practicability' is the one most frequently encountered in HSW regulatory requirements.

The ALARP suite of guidance, unlike R2P2, is intended primarily for HSE inspectors to help them judge whether duty holders have satisfactorily demonstrated that they have reduced risks to ALARP, that is, whether the decisions the duty holders made about reducing risks to ALARP were correct.

The ALARP suite of guidance consists of three documents covering:

1. **The basic principles and guidelines to be applied** – based on what the courts have said by way of interpreting the concept of 'reasonable practicability', that is, that it must

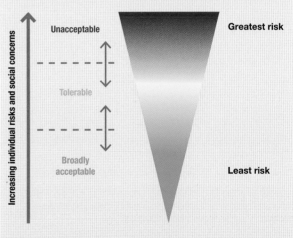

Figure 3.1 The tolerability of risk (TOR) framework

involve weighing the cost to the duty holder of instituting a control measure against the risk to be controlled, and ruling out the measure only if the cost is 'grossly disproportionate' to the risk. The document provides detailed guidance on issues such as what risks and costs should be taken into account in the trade-off and how should the trade-off be made.

2. **The role of 'good practice' in meeting ALARP** – *we expect it to be in place as a minimum requirement but it will often prove, in practice, to be sufficient in itself.*

3. **ALARP in design** – *considering ALARP at the design stage provides an opportunity to meet the very high potential for reducing risks throughout the life cycle of a hazard by designing in health and safety from the start, and allows consideration of any trade-off between the risk reductions to be achieved at different stages, in order to obtain the safest solution overall.*

Conclusion

Whilst acknowledging the limitations of the approach to decision-making provided by R2P2 and the ALARP guidance, we are confident that it has provided, and will continue to provide, a highly effective tool for both HSE and duty holders in deciding what to do about managing risk. We hope that the publication of R2P2 and the ALARP suite will help decision-makers reach decisions about managing risk in their own workplaces. The better they understand HSE's expectations, the more likely we are to both agree as to what needs to be done.

But with our own decision-making, we also need to understand the concerns of our stakeholders and society at large, and the values people employ when they consider matters of risk. We need to raise the public understanding of the issues involved. Prompting a more informed public debate on how to handle risk is an essential part of all this and we hope that publishing these documents will have helped to stimulate it.[9]

As Professor Löfstedt has said about this discussion of SFARP/ALARP, the 'tolerability of risk' and 'reasonably practicable' are, arguably, not sufficiently defined by the new Regulations and need further investigation and clarification to ensure shared understanding and consistent usage. However, the CONIAC draft Industry Guidance[34] for CDM 2015 goes much further to explain these relatively complex issues.

The General Principles of Prevention and examples of their application for CDM 2015

Rather than seeking to clarify tolerability of risk and SFARP, the CDM Regulations 2015 re-introduce the General Principles of Prevention as a 'precautionary principle' process for CDM integration. This is a target-setting checklist that designers must take into account when preparing or modifying a design. It is a framework for consideration of CDM issues rather than a strict hierarchy, and its language is still open to some considerable interpretation. These principles have been in the CDM Regulations and ACOP since 2007 and were also in the original EU Directive, but their significance was perhaps understated, so they were poorly applied.

The table below lists these principles and how you might apply them in practice. The first column in the table cites the General Principles of Prevention outlined in the EU Directive and CDM 2007, while the second column offers examples of applying them in practice as outlined in CDM 2015. The final column shows the author's comments and interpretations in accordance with SFARP.

Table 3.1 Examples of the General Principles of Prevention in CDM 2015

The proportionate application of these principles using SFARP is at the heart of CDM 2015.

	General principles of prevention (2007)	Examples of applying them in practice (2015)	Author's comments/ SFARP interpretations
A	Avoiding risks	Can I get rid of the problem (or hazard) altogether? For example: — can air conditioning plant on a roof be moved to ground level, so that work at height is not required for either installation or maintenance? — position a door away from a traffic route — design a roof with a high parapet to limit the risk of falls	While these are reasonable safety aspirations they may not be 'reasonably practicable' due to other factors, for example: — Plant on the roof is preferred for environmental and planning reasons. — Roof parapets can be expensive or visually obtrusive. They are not essential if other appropriate fall protection or prevention measures are provided.
B	Evaluating the risks which cannot be avoided	For example: work out whether the effort and expense of installing a fixed roof access system is appropriate if an area is only occasionally reached and can be done by using a MEWP	MEWP access is not always possible without a suitable space or hard standing, floor build-up specification or adjacent land ownership or access rights. But occasional inspections are necessary and so other types of suitable protection are required.
C	Combating the risks at source	For example: Arrange for services to be isolated and diverted to where they will be away from the work area	Services diversions can be expensive and dangerous in their own right. Both new services and existing installations require carefully planned routes on drawings in relation to other features such as piles or foundations.

	General principles of prevention (2007)	Examples of applying them in practice (2015)	Author's comments/ SFARP interpretations
D	Adapting the work to the individual, especially as regards the design of workplaces, the choice of work equipment and the choice of working and production methods, with a view, in particular, to alleviating monotonous work and work at a predetermined work rate and to reducing their effect on health	Providing workstations at an appropriate height, consult with workforce. For example: — position lighting so it can be accessed easily for maintenance such as positioning bulkhead lights on landings and not half way down staircases	Work positioning is the key to this issue. Find suitable locations to carry out plant maintenance, lamp replacement and cleaning activities and have appropriate work platforms to carry out the required activities.
E	Adapting to technical progress	Consider new techniques or technologies. For example: — self-cleaning glass — on tool dust extraction — roof inspections by drone	Self-cleaning glass is not always totally effective with heavy pollution and contamination, and itself requires replacement coatings at regular intervals. Technology can help but can also cause additional risks such as dust, noise and vibration.
F	Replacing the dangerous by the non dangerous or the less dangerous	Provide a less risky option. For example: — switch to using a paving-block that is lighter in weight — substitute solvent based products with water based equivalents — recycled tyre Kerbs instead of heavy concrete ones	The design intent may require: — larger paving slabs which require mechanised laying techniques; — solvent products which may have a longer life and require less frequent repainting; — granite kerbs, often preferred as rubber kerbs are not culturally acceptable in the UK yet.

	General principles of prevention (2007)	Examples of applying them in practice (2015)	Author's comments/ SFARP interpretations
G	Developing a coherent overall prevention policy which covers technology, organisation of work, working conditions, social relationships and the influence of factors relating to the working environment	Set standards. For example: Specify that all blocks should be cut using block splitter techniques rather than mechanical cutting which produces large amounts of harmful silica dust	Block splitters are not always appropriate for the various sizes of block available and the accuracy of cutting is variable. Pre-cut or pre-formed specials or other safe cutting systems such as benches with dust extraction measures may be more appropriate. Creating a level playing field for all contractors' pricing is essential.
H	Giving collective protective measures priority over individual protective measures	— Make provisions so that the work can be organised to reduce exposure to hazards (e.g. make provision for traffic routes so that barriers can be provided between pedestrians and traffic); — Fixed edge protection (barriers) rather than running lines	On-site segregation between people and vehicles is essential, but the interfaces need management, especially where users or the general public must be accommodated adjacent to the site. This is most applicable in the temporary condition on site but may not be appropriate for permanent situations.
I	Giving appropriate instructions to employees	Information on drawings or instructions such as intended sequencing	Providing site set-up, transient phases and significant residual risk information on drawings is a very powerful safety tool for all design and construction team employees.

Other health and safety frameworks

In parallel with the Principles of Prevention there are other health and safety 'frameworks' that permeate the education of health and safety practitioners particularly with respect to production industries but add to confusion at design stages as they tend to inhibit innovation or creativity.

> The first is the **hierarchy of hazard control** (see Fig. 3.2) – a system used in industry to minimize or eliminate exposure to hazards. It is a widely accepted system promoted by numerous safety organizations. This concept is taught to managers in industry, to be promoted as standard practice in the workplace.
>
> Here risks should be reduced to the lowest reasonably practicable level by taking preventative measures, in order of priority. This is what is meant by a hierarchy of control. The list below sets out the order to follow when planning to reduce risks you have identified in your workplace. Consider the headings in the order shown, do not simply jump to the easiest control measure to implement.
>
> 1. Elimination – Redesign the job or substitute a substance so that the hazard is removed or eliminated.

2. Substitution – Replace the material or process with a less hazardous one.
3. Engineering controls – for example use work equipment or other measures to prevent falls where you cannot avoid working at height, install or use additional machinery to control risks from dust or fumes or separate the hazard from operators by methods such as enclosing or guarding dangerous items of machinery/equipment. Give priority to measures which protect collectively over individual measures.
4. Administrative controls – These are all about identifying and implementing the procedures you need to work safely. For example: reducing the time workers are exposed to hazards (e.g. by job rotation); prohibiting use of mobile phones in hazardous areas; increasing safety signage, and performing risk assessments.
5. Personal protective clothes and equipment – Only after all the previous measures have been tried and found ineffective in controlling risks to a reasonably practicable level, must personal protective equipment (PPE) be used. For example, where you cannot eliminate the risk of a fall, use work equipment or other measures

continues on p32

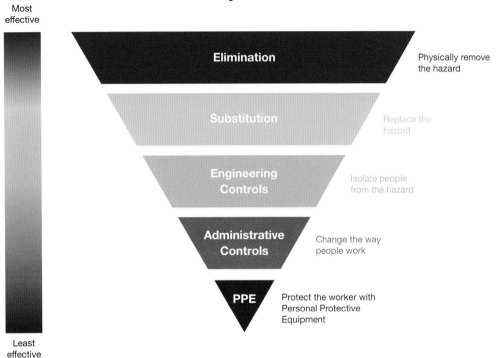

Figure 3.2 Hierarchy of Controls

> to minimise the distance and consequences of a fall (should one occur). If chosen, PPE should be selected and fitted by the person who uses it. Workers must be trained in the function and limitation of each item of PPE.[10]

A second example is the ERIC model introduced in the CDM 2007 Industry Guidance.

> This guidance illustrates the necessary sequence of actions by adopting the ERIC (an acronym for Eliminate, Reduce, Inform and Control) model because it is compatible with the need for a simple qualitative technique. It will only be on rare occasions (for example in the design of nuclear facilities) that a more sophisticated approach, such as quantitative analysis or risk ranking, will be required. There are a number of ways in which this can be achieved at a practical level and, correctly, CDM 2007 does not dictate a specific method.
>
> Eliminate Reduce Inform – Control
>
> The ERIC acronym indicates the required actions and the required sequence of the design risk assessment process. The separating line before the 'C' emphasises that once a designer has passed on appropriate information (to the contractor, those undertaking the maintenance or to the user), then providing the design does not change (directly or through the impact of an adjacent influence, for example) the Control of the resulting risks then belongs to other duty holders. Only after having considered one 'letter' in the ERIC acronym does a designer move on.[11]

While these apparently simple and logical processes apply easily to the production, construction and engineering industries, they have been misunderstood and misapplied in the architectural and creative design industries. The concept of 'elimination' of hazards can lead to the idea that an innovative or unusual design constitutes a hazard that needs elimination. Hence misinformed health and safety practitioners require the elimination of the design intent element, feature or building form itself rather than the risks associated with the design. This attitude in turn leads to an atmosphere of confrontation rather than collaboration – where the desirable outcome would be for designers to feel able to put forward innovative and eclectic solutions, managed within an appropriate framework of consideration.

There is therefore still a considerable need to develop understanding of these three types of precautionary principles in relation to creative design development and the interpretation of SFARP.

4 CDM 2007 and CDM 2015 compared and contrasted

Chapter overview

- A comparison between CDM 2007 and 2015
- The CDM 2015 duty holders
- CDM 2015 flowchart
- Identifying the key duty holder changes from CDM 2007 to 2015
- Comparing competence in CDM 2007 and 2015

A comparison between CDM 2007 and 2015

The objectives of CDM 2007 have been generally retained, improved where not working as intended and clarified in a more concise set of documentation in the new 2015 Regulations. The Industry Guidance[34] is less complicated and more accessible than in 2007, and should help to embed the changes in all areas of the industry from the largest project to the smallest. While it has been written with small and medium enterprises in mind, it is applicable to all sizes of projects.

In this chapter the 2007 and 2015 Regulations and Industry Guidance, at the time of writing, are compared to identify how they have changed in general terms. A CDM 2015 flowchart (Fig. 4.1) explains how the roles of the duty holders work together with their aims and responsibilities, and how this is different from CDM 2007. Finally there is comparison between how CDM 2007 and 2015 define 'competence' in how duty holders' responsibilities are carried out. The draft issued in January 2015 is subject to further consultation; a further revision is due to come into force in April 2015.

Table 4.1 An overview of CDM 2007 and 2015

General issues in CDM 2007	General issues in CDM 2015
The CDM Regulations were intended to ensure that health and safety issues were properly considered during a project's development to reduce the risk of harm to those who have to build, use and maintain structures. The Approved Code of Practice, Managing health and safety in construction, provides practical guidance about how to comply with the Regulations, including detailed descriptions of the duties imposed by them. The Code proposes that: '... The key aim of CDM 2007 is to integrate health and safety into the management of the project and to encourage everyone involved to work together to: a) improve the planning and management of projects from the very start; b) identify hazards early on, so they can be eliminated or reduced at the design or planning stage and the remaining risks can be properly managed; c) target effort where it can do the most good in terms of health and safety; d) discourage unnecessary bureaucracy.'[3]	It is proposed that a further revision will come into force in April 2015. The HSE's *CD261 – Consultation on replacement of the Construction (Design and Management) Regulations 2007* suggests that: 'There has been significant improvement in the industry's performance on health and safety over recent years. However, it remains one of the highest risk industry sectors in which to work – with unacceptable standards, particularly on smaller sites ...' **Key elements to securing construction health and safety** These are: a) managing the risks to health and safety by applying the general principles of prevention; b) **appointing** the right people and organisations at the right time; c) making sure everyone has the **information, instruction, training and supervision** they need to carry out their jobs in a way that secures health and safety; d) duty holders **cooperating and communicating** with each other and **coordinating** their work; and e) **consulting workers and engaging** with them to promote and develop effective measures to secure health, safety and welfare.

General issues in CDM 2007

The Regulations were divided into five parts:
Part 1 deals with matters of interpretation and application of the Regulations.
Part 2 covers general management duties of the duty holders that apply to all construction projects including those that are non-notifiable.
Part 3 sets out additional management duties of the duty holders that apply to notifiable projects.
Part 4 applies to all construction work carried out on construction sites, and covers physical safeguards that must be provided to prevent danger. This was covered previously by the Construction (Health, Safety and Welfare) Regulations 1996, which were revoked by CDM 2007.
Part 5 covers issues of civil liability, transitional provisions that apply during the period when the regulations come into force, and amendments and revocations of other legislation.

CDM 2007 applies to all construction work and covers a very broad range of construction activities, such as building, civil engineering, engineering construction work, demolition, site preparation and site clearance – except for Part 3, which only applies if the project is notifiable. On a notifiable project, the client must additionally appoint a competent CDM coordinator and a competent principal contractor, a construction phase plan – incorporating mobilisation (let the building contract, appoint contractor, issue production information,

General issues in CDM 2015

The draft Regulations therefore propose the following changes:
1. **Structural simplification** of the Regulations to make them easier to understand.
2. **The Approved Code of Practice (ACOP)** will be signposting only with more targeted guidance.
3. **CDM coordinator** role is removed and new role of principal designer is introduced.
4. **Competence requirements** are replaced with a specific requirement for appropriate skills.
5. **Domestic clients'** exemption is removed, but their CDM duties are passed to the contractor.
6. **Threshold for appointment** of coordinators: this will be changed to require coordinators where there is more than one contractor. The HSE suggests that this will capture close to an additional 1 million projects per year, but that the requirements will be proportionate and little more work will be necessary than at present. Some concern has been expressed about what constitutes more than one contractor, and how it is possible to know how many contractors may be needed. In addition, it will separate the threshold for coordination from that of notification.

In terms of the organisation of projects, the most significant of the changes is the removal of the CDM coordinator role and the introduction of the principal designer role (PD.) The reason for the change is to give responsibility for CDM during the design phase

General issues in CDM 2007	General issues in CDM 2015
arrange site handover and review contractor's proposals), construction to practical completion (administer the building contract and provide contractor with further information as necessary), and after practical completion (administer the building contract after practical completion, resolve defects and make final inspections). The importance of this phase is in the inspection of expected performance standards to ensure compliance), and a health and safety file must be produced. Additional duties are also placed on the clients, designers and contractors for notifiable projects. The CDM coordinator is the new title for the planning supervisor under CDM 1994, with increased duties and responsibilities.	to an individual who has the ability to influence the design. At present, this role is often contracted out, resulting in extra costs, but the individual appointed is rarely properly embedded in the project team and so has little opportunity to influence the design. The role of principal designer is analogous to that of the principal contractor during the construction phase. This will be a significant change and may require amendment of appointment documents and contracts. It is likely that consultants currently acting as CDM coordinators with design expertise will in future act as client advisors or assist principal designers with their new role. Principal designers will generally have the requisite coordination skills for integration of CDM, but will have to upgrade their specialist health and safety and CDM knowledge to carry out the full role. As a result of the public consultation a much-simplified ACOP may be produced in 2016, and a six-month transitional period will be implemented in 2015.

The following section is a summary of the duty holders' responsibilities and a comparison schedule for CDM 2007 and 2015 to explain the changes to these responsibilities. There is also commentary from the author to explain what the key changes are and why they have been made. These are summarising extracts from the HSE 2015 consultation documents, the draft *(L) Series guidance*[8] and the CITB draft Industry

Guidance for Principal Designers[34] issued in January 2015. While every effort has been made to capture the key issues in this document, certain details will be missed and there will be further revisions by the time of publication in April 2015. Readers should acquaint themselves with the general principles here and then refer to the completed primary source documents.

The CDM 2015 duty holders

Table 4.2 Duty holders and their roles summarised in CDM 2015

CDM 2015 Duty holders* – Who are they?	Summary of role and main duties What do they do?	Author's comments and observations
Clients Organisations or individuals for whom a construction project is carried out.	Make suitable arrangements for managing a project. This includes making sure that: — other duty holders are appointed — sufficient time and resources are allocated. Clients must also make sure that: — relevant information is prepared and provided to other duty holders — the principal designer and principal contractor carry out their duties — welfare facilities are provided.	The client duties have largely remained and been extended but are explained more clearly in 2015 and in a more appropriate format. Some additional duties arise from the removal of the CDM-C role, e.g. notification of projects and checking that the team is adequately resourced.
Domestic clients People who have construction work carried out on their own home or the home of a family member, that is not done in furtherance of a business, whether for profit or not.	Domestic clients are in scope of CDM 2015, their duties as a client are normally transferred to: — the contractor on a single contractor project, or — the principal contractor on a project involving more than one contractor. However the domestic client may can choose to have a written agreement with the principal designer to carry out the client duties	Domestic clients cannot be expected to exercise the duties of commercial clients. By transferring these duties to their consultant team the EU Directive is met and an appropriate level of design coordination and site safety are integrated into all projects.

CDM 2015 Duty holders* – Who are they?	Summary of role and main duties What do they do?	Author's comments and observations
Principal designers** Designers appointed by the client where projects involve more than one contractor. They can be an individual or an organisation with the sufficient knowledge, experience and ability to carry out the role.	Plan, manage, monitor and coordinate health and safety in the pre-construction phase of a project. This includes: — identifying, eliminating or controlling foreseeable risks — ensuring designers carry out their duties. Prepare and provide relevant information to other duty holders. Liaise with the principal contractor to help in the planning, management, monitoring and co-ordination of the construction phase.	Principal designers will need to upskill (or engage suitable support) to build an awareness of CDM issues, but the integration of CDM into designs is primarily about the consideration of all other factors in relation to health and safety (SFARP).
Designers Those who, as part of a business, prepare or modify designs for a building, product or system relating to construction work.	When preparing or modifying designs eliminate, reduce or control foreseeable risks that may arise during: — construction, and — the maintenance and use of a building once it is built. — Providing information to other members of the project team to help them fulfil their duties.	These duties are essentially the same as before, but better explained.
Principal contractors Contractors appointed by the client to coordinate the construction phase of a project where it involves more than one contractor.	Plan, manage, monitor and co-ordinate the construction phase of a project. This includes: — liaising with the client and principal designer — preparing the construction phase plan — organising co-operation between contractors and co-ordinating their work.	The PC role is largely the same as before but with a greater level of responsibility to embed good H&S attitudes into the smaller end of the construction industry, as well as health issues throughout.

CDM 2015 Duty holders* – Who are they?	Summary of role and main duties What do they do?	Author's comments and observations
The principal contractor is normally a contractor so will also have contractor duties.	Ensure that: — suitable site inductions are provided — reasonable steps are taken to prevent unauthorised access — workers are consulted and engaged in securing their health and safety — welfare facilities are provided.	Essentially the same as 2007 but relates to all sites with more than one contractor.
Contractors Those who do the actual construction work and can be either an individual or a company. A contractor may be an individual, a sole trader, a self-employed worker or a business who carries out, manages or controls construction work in connection with a business.	Plan, manage and monitor construction work under their control so that it is carried out without risks to health and safety. For projects involving more than one contractor, co-ordinate their activities with others in the project team – in particular, comply with directions given to them by the principal designer or principal contractor. For single-contractor projects, prepare a construction phase plan.	Contractors' duties are largely as before but better explained, and more appropriate for all levels of the industry.

CDM 2015 Duty holders* – Who are they?	Summary of role and main duties What do they do?	Author's comments and observations
Workers The people who work for or under the control of contractors on a construction site.	They must: — be consulted about matters which affect their health, safety and welfare — take care of their own health and safety and that of others who may be affected by their actions — report anything they see which is likely to endanger either their own or others' health and safety — co-operate with their employer, fellow workers, contractors and other duty holders.	Workers are not duty holders but there are still specific expectations placed on them.

Note

*   Organisations or individuals can carry out the role of more than one duty holder, provided they have the skills, knowledge, experience and (if an organisation) the organisational capability necessary to carry out those roles in a way that ensures health and safety.

** Principal designers now take on some of the responsibilities of the defunct CDM coordinator role.

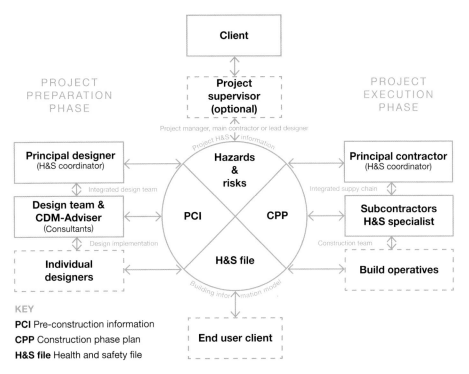

Figure 4.1 CDM 2015 flowchart
This flowchart shows the now simple and balanced structure of the Project Preparation (Design) and the Project Execution (Construction) phases of projects under CDM 2015 as originally intended by the EU Directive. The central zone is the project information that needs to be collectively assembled and communicated throughout the project.

Note: The project supervisor is an optional appointment under the EU Directive and is common in the UK industry as a 'project manager'. They should be recognised with CDM responsibilities, even though not mentioned in CDM 2015.

Identifying the key duty holder changes from CDM 2007 to 2015

CDM Regulations 2015 for clients

The CDM Regulations involve a major change for clients. Client duties are now extended, and the new function of principal designer (taking over some of the responsibilities of the now defunct CDM coordinator) will have limited involvement with the client. The client therefore loses their main CDM 'advisor', but may choose to appoint one independently.

Key changes include:
1. CDM coordinator removed. New principal designer role created from within the design team.
2. Domestic client exemption removed and domestic client CDM duties transferred to the designer or contractor.
3. CDM competency requirements removed.
4. Approved Code of Practice (ACOP) replaced with a 'signposting ACOP', *(L) Series guidance*[8] and *Industry Guidance*.[34]

There is a six-month transitional period from 6 April 2015 (for projects in construction) to re-appoint the team to fulfil the new roles.

Table 4.3 A comparison of client duties in CDM 2007 and 2015

Client CDM 2007 (ACOP)	Client CDM 2015 (Regulations & Guidance)
Client duties The client has one of the biggest influences over the way a project is run. They have substantial influence and contractual control, and their decisions and approach determine: — the time, money and other resources available for projects; — who makes up the project team, their competence, when they are appointed, and who does what; — whether the team members are encouraged to cooperate and work together effectively; — whether the team has the information that it needs about the site and any existing structures; — the arrangements for managing and coordinating the work of the team. Because of this, clients are made accountable for the impact their approach has on the health and safety of those working on or affected by the project. However, the Regulations also recognise that many clients know little about construction health and safety, so clients are not required or expected to plan or manage projects themselves. Nor do they have to develop substantial expertise in construction health and safety, unless this is central to their business.	**Client duties** The client has the overall responsibility for the successful execution of the project. The client has contractual control, and your influence, decisions and approach can determine: — the time, money and other resources available for the project; — who makes up the project team, their level of knowledge, skills and experience, and their roles and responsibilities; — the arrangements for managing and coordinating the work of the project team that has been assembled to carry out the project; — how members of the project team work together; — the adequacy of information available at the planning or pre-construction phase; — the style, tone and culture of communications used throughout the project. The client is supported by the principal designer and principal contractor for different phases of the project. As part of this support, the principal designer and the principal contractor have an important role in coordinating health and safety. It is therefore essential that these three duty holders must have good working relationships from the outset.

Client CDM 2007 (ACOP)	Client CDM 2015 (Regulations & Guidance)

Clients must ensure that various things are done, but are not normally expected to do them themselves.
 	
In the case of notifiable projects, clients must appoint a competent CDM coordinator. Those clients without construction expertise should rely on the CDM coordinator's advice on how best to meet their duties, but the CDM coordinator will need the client's support and input to be able to carry out their work effectively. The client remains responsible for ensuring that client duties are met.
 	
Clients must take reasonable steps to make sure arrangements for managing the project by duty holders are suitable, to ensure the following:
— Construction work can be carried out so far as is reasonably practicable without risk to health and safety.
— Suitable welfare facilities are provided for any person carrying out the construction work.
— Any structure designed for use as a workplace has been designed taking into account the Workplace (Health, Safety and Welfare) Regulations 1992 (the design of, and materials used in, the structure).
— Pre-construction information is provided promptly to every person designing the structure and every contractor who has been or may be appointed by the client.
	This enables the provision and flow of information to ensure that health and safety is considered when making decisions.
	
	This is the arrangement for the majority of projects. The only exception is when the client does not need to appoint a principal designer or principal contractor because the work is to be undertaken by a single contractor.
	
	Clients are organisations or individuals for whom a construction project is carried out.
	— The role of the client is crucial in ensuring that a construction project is carried out from start to finish in a way that adequately controls the risks to the health and safety of those who may be affected.
	— Their main duty is to make suitable arrangements for managing a project to secure health, safety and welfare throughout its life.
	— The domestic client is in the scope of CDM 2015, but their duties as a client are normally transferred to either:
	  — the contractor on a single contractor project; or
	  — the principal contractor when there is more than one contractor on the project.
	The domestic client may however choose to appoint a principal designer and enter into a written agreement for the principal designer to carry out the client duties.

Client CDM 2007 (ACOP)

The client's role
Where projects are notifiable under the Regulations (projects which last more than 30 days or involve more than 500 person days of construction work), the client must also:

— appoint a CDM coordinator;
— appoint a principal contractor;
— ensure construction work does not start unless a construction phase plan is in place;
— ensure construction work does not start unless there are adequate welfare facilities on site;
— provide information relating to the health and safety file to the CDM coordinator;
— retain and provide access to the health and safety file.

At present, domestic clients do **not** normally have any duties under the Regulations (although they may have duties if they control the way the work is carried out).

Client CDM 2015 (Regulations & Guidance)

The client's role
The client can fulfil their role in a number of ways before, during and after each stage of the project. Under CDM 2015 their duties are summarised below.

Before building work begins (design or pre-construction stage)
— Make suitable arrangements for managing the project.
— Select the project team and formally appoint duty holders.
— Provide information to help with design and construction planning.
— Notify the project to the enforcing authorities (where required).
— Check that the principal designer is carrying out their duties.
— Check that all pre-construction activities have been carried out and the construction phase plan drawn up.
— Ensure that suitable welfare facilities are in place.

Domestic clients are people who have construction work carried out on their own home, or the home of a family member that is not for business purposes, whether for profit or not.

Client CDM 2007 (ACOP)	Client CDM 2015 (Regulations & Guidance)
Clients are **not** required or expected to: — plan or manage construction projects themselves; — specify how work must be done, for example requiring a structure to be demolished by hand. Indeed they should not do so unless they have the expertise to assess the various options and risks involved. (They should, of course, point out particular risks that would inform this decision); — provide welfare facilities for those carrying out construction work (though they should cooperate with the contractor to assist with their arrangements); — check designs to make sure that Regulation 11 has been complied with; — visit the site (to supervise or check construction work standards); — employ third-party assurance advisors to monitor health and safety standards on site (though there may be benefits to the client in doing so); — subscribe to third-party competence assessment schemes (though there may be benefits in doing so).	The domestic client is in the scope of CDM 2015, but their duties as a client are normally transferred to either: — the contractor on a single-contractor project; or — the principal contractor when there is more than one contractor on the project. The domestic client may however choose to appoint a principal designer and enter into a written agreement for the principal designer to carry out the client duties. The HSE wants to ensure that effective coordination of health and safety is carried out on all projects regardless of whether they are carried out for domestic clients. It is also clear that it expects these requirements to be discharged in a sensible and proportionate manner. For the majority of small, domestic projects this will mean no change to how these projects should currently be managed for health and safety. As part of the proposed guidance the HSE will make this clear, as well as the associated enforcement expectations. **During the building work (construction stage)** — Ensure the construction phase plan is in place. — Ensure welfare facilities are in place. — Ensure that the arrangements for managing a project are working. — Check that the principal contractor is carrying out their duties. — Check completion and handover arrangements.

Client CDM 2007 (ACOP)	Client CDM 2015 (Regulations & Guidance)
	After building work has finished (post-construction stage) — Check that the principal designer has prepared the health and safety file, or the principal contractor where the principal designer's role has finished. — Keep the health and safety file available for future work on the structure, and for any new owners.

Table 4.4 A comparison of planning supervisor and CDM-C duties in CDM 2007 with principal designer duties in CDM 2015

Planning Supervisor 1994 & CDM-C 2007	Principal Designer CDM 2015
Planning supervisor (CDM 1994) The role of 'planning supervisor' was created in 1994. The Regulations were introduced to ensure that health and safety issues were properly considered during a project's development so that the risk of harm to those who have to build, use and maintain structures was reduced. The planning supervisor role was intended to be a supervisory one, ensuring that the Regulations were complied with by: — ensuring designers cooperated, and avoided and reduced risks; — ensuring that the health and safety plan and file were prepared; — advising the client about the adequacy and competence of designers and contractors; — advising the client about the health and safety plan; — ensuring that the HSE was notified of the project. **The role of CDM coordinator (CDM 2007) on notifiable projects only** When the CDM Regulations were revised in 2007, planning supervisors were replaced with a new role of CDM coordinator (CDM-C). The CDM-C was intended to take a more active role in the project, actually coordinating health and safety issues and performing tasks themselves rather than just advising or ensuring that others perform their duties.	**Who should be the principal designer?** The principal designer must be a designer on the project and must be in a position to have control over the design and planning stage. **A designer is an organisation or individual that prepares or modifies a design for a construction project, including the design of temporary works, or arranges for or instructs someone else to do so.** The principal designer will usually be an organisation or, on smaller projects, an individual with: — a technical knowledge of the construction industry, relevant to the project; — an understanding of how health and safety is managed through the design process; — the skills to be able to oversee health and safety during the pre-construction phase of the project and the ongoing design. **Appointing the principal designer** You must be appointed in writing by the client. You can combine your role as principal designer with other roles, such as the project manager. This will assist with the integration of health and safety into the project.

Planning Supervisor 1994 & CDM-C 2007

What CDM coordinators 2007 should do
- Give suitable and sufficient advice and assistance to clients in order to help them to comply with their duties, in particular:
  - the duty to appoint competent designers and contractors; and
  - the duty to ensure that adequate arrangements are in place for managing the project.
- Notify the HSE about the project – submit the F10.
- Coordinate design work, planning and other preparation for construction where relevant to health and safety.
- Identify and collect the pre-construction information and advise the client if surveys need to be commissioned to fill significant gaps.
- Promptly provide in a convenient form to those involved with the design of the structure, and to every contractor (including the principal contractor) who may be or has been appointed by the client, the parts of the pre-construction information that are relevant to each.
- Manage the flow of health and safety information between clients, designers and contractors.
- Advise the client on the suitability of the initial construction phase plan and the arrangements made to ensure that welfare facilities are on site from the start.

Principal Designer CDM 2015

The principal designer's role
The principal designer's role is to plan, manage and monitor the coordination of the pre-construction phase, including any preparatory work carried out for the project. You must first ensure that the client is aware of their duties, and:
- assist the client in identifying, obtaining and collating the pre-construction information;
- provide the pre-construction information to all designers, the principal contractor and contractors;
- ensure that designers comply with their duties;
- liaise with the principal contractor for the duration of your appointment;
- prepare the health and safety file.

What must a principal designer do?
They must identify, eliminate or control foreseeable risks.

Principal designers must ensure, so far as is reasonably practicable, that foreseeable risks to health and safety are identified (Regulation 11(3)). In practice, this will involve the principal designer working with other designers involved with the project. The risks that should be identified are those that are significant and are likely to arise:
- while carrying out construction work; or
- during maintenance, cleaning or the use of the building as a workplace once it is built.

Planning Supervisor 1994 & CDM-C 2007

— Produce or update a relevant, user-friendly health and safety file suitable for future use at the end of the construction phase.

What CDM coordinators don't have to do
— Approve the appointment of designers, principal contractors or contractors, although they normally advise clients about competence and resources.
— Approve or check designs, although they have to be satisfied that the design process addresses the need to eliminate hazards and control risks.
— Approve the principal contractor's construction phase plan, although they have to be able to advise clients on its adequacy at the start of construction.
— Supervise the principal contractor's implementation of the construction phase plan – this is the responsibility of the principal contractor, or supervise or monitor construction work – this is the responsibility of the principal contractor.

Replacing the CDM coordinator with the principal designer role – on projects with more than one contractor

The HSE proposes to remove the CDM-C role. There is a widely held view, supported by evidence, that the current approach is often bureaucratic and adds costs with little added value. As such it is often ineffective.

Principal Designer CDM 2015

Once the risks have been identified, principal designers must follow the approach to managing them set out in the General Principles of Prevention. The principal designer must, so far as reasonably practicable, ensure that the design team:
— **eliminate** the risks associated with design elements. If this is not possible (for instance because of competing design considerations such as planning restrictions, specifications, disproportionate costs or aesthetics);
— **reduce** any remaining risks; or
— **control** them, to an acceptable level. This relies on exercising professional judgement in considering how the risks can be managed. The focus should be on those design elements where there is a significant risk of injury or ill-health.

It is not the principal designer's responsibility to:
— submit the notification (F10) to the HSE or check that the client has done so;
— check the skills and experience of the designers or contractors unless you are appointing them directly;
— advise the client on their appointment of designers and contractors, including their skills and experience;
— advise the client on their health and safety arrangements for the project, including welfare facilities;

Planning Supervisor 1994 & CDM-C 2007

Appointments are often made too late, too little resource is made available, and those involved in fulfilling the role are often not well embedded into the pre-construction project team.

Pre-construction coordination is required by the EU Temporary and Mobile Construction Sites Directive. The aim is to replace the role with that of 'principal designer' (PD). The responsibility for discharging the function will rest with an individual or business in control of the pre-construction phase. It is this element of control and influence over the design that is the fundamental difference between the CDM-C role and the PD role. The default position is that the responsibility for this function will come from within the existing project team, facilitating an integrated approach to risk management. The HSE expects that moving away from a default position where an external contractor is appointed will deliver considerable economies of scale.

Principal Designer CDM 2015

— review or approve the construction phase plan or check that it has been implemented;
— review or approve health and safety arrangements on site, including method statements;
— take on overall responsibility for the project – your role is only to manage health and safety during the pre-construction phase;
— supervise or monitor health and safety on site – this is the responsibility of the principal contractor;
— check or approve designs other than when reviewing health and safety risks.

The principal designer may wish to offer any of the activities listed above as additional services.

There could be some misconceptions around the role of the principal designer, especially as it is not a direct replacement for the CDM-C as defined in the CDM Regulations 2007.

There are more duties than a designer needs to execute, but fewer than the 2007 CDM-C was intended to carry out. However it is important to identify the additional duties and the costs associated in fee proposals in order to provide adequate resources.

Table 4.5 A comparison of designer duties in CDM 2007 and 2015

Designer CDM 2007	Designer CDM 2015
Who is a designer? The Regulations define a designer as '… any person (including a client, contractor or other person referred to in these Regulations) who in the course or furtherance of a business: — prepares or modifies a design; or — arranges for or instructs any person under his control to do so, relating to a structure or to a product or mechanical or electrical system intended for a particular structure, and a person is deemed to prepare a design where a design is prepared by a person under his control.' **Where 'design' includes** '… drawings, design details, specification and bill of quantities (including specification of articles or substances) relating to a structure, and calculations prepared for the purpose of a design.' **Designers should:** — make sure that they are competent and adequately resourced to address the health and safety issues likely to be involved in the design; — check that clients are aware of their duties;	**Who is a designer?** A designer is an organisation or individual who prepares or modifies a design for any part of a construction project, including the design of temporary works, or arranges for or instructs someone else to do so. Designers can include architects, consulting engineers, quantity surveyors, interior designers, temporary work engineers, chartered surveyors, technicians, specifiers, principal contractors, specialist contractors and some tradespeople. There are other specialist supply-chain personnel who could also be described as designers (see Industry Guidance).[34] **What is a design?** Design includes drawings, sketches, design details, specifications and product selection, bills of quantity or calculations, prepared for the purpose of constructing, modifying or using a building or structure, a product or a system (such as a mechanical or electrical system). **Designers must:** — understand and be aware of significant risks that construction workers and users can be exposed to, and how these can arise from design decisions; — have the right skills, knowledge and experience, and be adequately resourced to address the health and safety issues likely to be involved in the design;

Designer CDM 2007	Designer CDM 2015
— when carrying out design work, avoid foreseeable risks to those involved in the construction and future use of the structure, and in doing so, they should eliminate hazards (so far as is reasonably practicable, taking account of other design considerations) and reduce risks associated with those hazards which remain; — provide adequate information about any significant risks associated with the design; — coordinate their work with that of others in order to improve the way in which risks are managed and controlled. In carrying out these duties, designers need to consider the hazards and risks to those who: — carry out construction work, including demolition; — clean any window or transparent or translucent wall, ceiling or roof in or on a structure or maintain the permanent fixtures and fittings; — use a structure designed as a place of work; — may be affected by such work, for example customers or the general public.	— check that clients are aware of their duties; — cooperate with others who have responsibilities, in particular the principal designer; — take into account the general principles of prevention when carrying out your design work; — provide information about the risks arising from their design; — coordinate their work with that of others in order to improve the way in which risks are managed and controlled. **Being appointed and appointing others** You and anyone you engage to help you with a design must have the appropriate skills, knowledge and experience to do the work. You may be asked to demonstrate this by providing simple evidence, such as proof of membership of a professional institution, references from previous clients or by showing examples of past work and how you have managed risks on similar projects. As a designer you should aim to eliminate risks, or else reduce them and inform those who need to know of the significant remaining risks. You will have to balance a wide range of interests, from client and user requirements; to planning constraints and aesthetics; to third-party, ecological, environmental, heritage or legal requirements.

Designer CDM 2007

The duties of designers include:
- Ensuring **the client is aware** of their duties under the regulations.
- **So far as is reasonably practicable, avoiding foreseeable** risks to the health and safety of any person that is:
  - carrying out construction work;
  - liable to be affected by such construction work;
  - cleaning any window or any transparent or translucent wall, ceiling or roof in or on a structure;
  - maintaining the permanent fixtures and fittings of a structure; or using a structure designed as a workplace.
- **Eliminating hazards** which may give rise to risks.
- **Reducing** risks from any remaining hazards.
- Giving **collective measures** priority over individual measures.
- Taking account of the provisions of the Workplace (Health, Safety and Welfare) Regulations 1992 which relate to the design of, and materials used in, the structure.
- Taking all reasonable steps to provide **sufficient information** to the client, other designers and contractors.

Designer CDM 2015

What do I have to do?
- **Making clients aware of their duties**
  When the client engages you to carry out design work you must make sure that they understand their responsibilities under CDM 2015 before you start. If the client needs more details about their responsibilities, the designer should refer them to the CONIAC Industry Guidance for Clients (CDM15/1). On projects with more than one contractor the client must appoint a principal designer. If you are working as one of a team of designers, it is important that you know who the principal designer is, and that you cooperate with them.
- **Eliminating, reducing and controlling risks through design**
  - As a designer you will need to take account of the General Principles of Prevention when preparing or modifying your design. The principles provide a framework within which a design is considered for any potential health and safety risks.
  - Health and safety risks must be considered alongside other factors that influence the design, such as cost, fitness for purpose, aesthetics and environmental impact.
  - When considering health and safety risks, you are expected to do what is reasonable at the time that the design is prepared, taking into account current industry knowledge and practice.

Designer CDM 2007	Designer CDM 2015
Where projects are notifiable under the Regulations (projects which last more than 30 days or involve more than 500 person days of construction work), the additional duties of the designer include: — **Ensuring a CDM coordinator has been appointed** for the project. NB It is generally accepted that 'design' commences at concept design stage, so this should not begin until a CDM-C is in place. — **Taking all reasonable steps to provide sufficient information** to assist the CDM-C to fulfil their duties, including those in relation to the health and safety file.	— Risks that cannot be addressed at the initial stage of a project should be reviewed later on, during the detailed design stage. — You should take into account the requirement for maintenance, cleaning and access to the finished project. Discussing this with those who will be carrying out this work is important. They may have established methods of working or specific needs or suggestions which you will need to consider in your design. — The level of detail required in passing on information about risks should be proportionate to the risks involved. Insignificant risks can usually be ignored, as can risks arising from routine construction activities, unless the design compounds or significantly alters these risks. — You could offer suggestions for inclusion in the pre-construction information about how elements of the final structure can be utilised during the construction phase, for example by installing the permanent stairs early in the build to reduce the need for scaffolding or temporary access. This will not only have health and safety benefits but could also reduce the project's overall time and cost. — Any records you wish to keep should not be overcomplicated, but proportionate to the risks involved as a reminder of why decisions were made. Examples of what to record include

Designer CDM 2007	Designer CDM 2015
	minutes or notes of meetings, notes on drawings and sketches, as well as risk registers and similar items on more complex projects. — If you are unsure how the design might be constructed, or are not aware of certain construction or maintenance techniques, talk to possible contractors, specialists, manufacturers or suppliers before completing your design. — **Preparing or modifying designs for safety and health** — Designers can help avoid and reduce the risks that arise during construction and associated work. — When preparing or modifying designs, your first aim is to eliminate risks to anyone who may be affected by your design or, if that is not possible, to reduce or control the risks. — Design is rarely a simple one-step operation. It usually involves you making changes as a result of discussion with others and as more information becomes available. Your design may also become more detailed as the project goes from concept to fully detailed proposals. — Your design will require you to apply your professional or trade expertise to produce information needed by others. They will be relying on you to do this so you should make sure that the information can be clearly understood by those who will use it.

Designer CDM 2007	Designer CDM 2015

	— **Cooperation and coordination**
	  — You must cooperate with the client, other designers and anyone else who will make use of your design or provides you with information, in particular the principal designer. This includes temporary and permanent works designers, who should themselves cooperate to ensure that their designs are compatible with each other.
	  — Depending on the nature and extent of design work, there may be a need to carry out design reviews in order to focus on areas of the design where there are health and safety risks requiring resolution.
	  — On projects where more than one contractor is involved, the principal designer should take the lead in managing this review process. For example, they may ask you to review your design when a subsequent designer or contractor asks for a change. On smaller projects these reviews could be part of normal project meetings.
	  — Reviews enable the project team to focus specifically on health and safety matters. They are most effective when held at the earliest opportunity so that risks can be identified and then eliminated or reduced in good time. The need for such reviews is likely to continue throughout the project. This is particularly necessary where there are changes to requirements or designs later in the project.

Table 4.6 A comparison of principal contractor duties in CDM 2007 and 2015

Principal Contractor CDM 2007	Principal Contractor CDM 2015
Additional duties for notifiable projects Note: There was no principal contractor on non-notifiable projects.	**The role of principal contractor** This applies to all projects with two or more contractors. The term **manage** in this guide means **plan**, **manage**, **monitor** and **coordinate** the construction phase so that health and safety risks are controlled. Key actions include:
— Plan, manage and monitor the construction phase in liaison with the contractor. — Prepare, develop and implement a written plan and site rules. (The initial plan should be completed before the construction phase begins.) — Give contractors relevant parts of the plan. — Make sure suitable welfare facilities are provided from the start and maintained throughout the construction phase. — Check the competence of all appointees. — Ensure all workers have site inductions and any further information and training needed for the work. — Consult with the workers. — Liaise with the CDM coordinator about the ongoing design. Secure the site.	— **planning** – preparing a construction phase plan that ensures the work is carried out without risk to health or safety; — **managing** – implementing the plan, including facilitating cooperation and coordination between contractors; — **monitoring** – reviewing, revising and refining the plan and checking work is being carried out safely and without risks to health; — **securing the site** – taking steps to prevent unauthorised access to the site by using fencing and other controls; — **providing welfare facilities** – making sure that facilities are provided throughout the construction phase; — **providing site induction** – giving workers, visitors and others information about risks and rules that are relevant to the site work and their work; — **liaising on design** – discussing with the principal designer any design or change to a design. The effort they devote to carrying out their responsibilities should be in proportion to the size and complexity of the site and the range and nature of the health and safety risks involved.

Principal Contractor CDM 2007

What principal contractors must do
Principal contractors must:

— satisfy themselves that clients are aware of their duties, that a CDM-C has been appointed and the HSE has been notified before they start work;

— make sure that they are competent to address the health and safety issues likely to be involved in the management of the construction phase;

— ensure that the construction phase is properly planned, managed and monitored, with adequately resourced, competent site management appropriate to the risk and activity;

— ensure that every contractor who will work on the project is informed of the minimum amount of time that they will be allowed for planning and preparation before they begin work on site;

— ensure that all contractors are provided with the information about the project that they need to enable them to carry out their work safely and without risk to health. Requests from contractors for information should be met promptly;

— ensure safe working and coordination and cooperation between contractors;

— ensure that a suitable construction phase plan ('the plan') is:

Principal Contractor CDM 2015

The principal contractor must carry out certain duties to fulfil their role.

— **Keep everyone healthy and safe on site**
You must consider how you will pass on any information you have received from the client and the designer to everyone else on site. This isn't a one-off task, as things may change during the project.

— **Ensure all aspects of health and safety are planned**
Planning is particularly important when work involves such issues as asbestos, lifting, demolition, excavation, working at height, structural work, electrical or confined space work.

— **Consider supervision requirements**
You must assess the degree of supervision you will need, taking account of the training, experience and likely behaviour of your workers. You should make arrangements to provide an adequate number of supervisors who have the necessary training, experience and leadership qualities for the risks that the project is likely to involve.

— **Ensure everyone receives the appropriate information**
Make sure that the work is organised properly and everybody has the information they need before commencing work. Let everyone coming on to the

Principal Contractor CDM 2007	Principal Contractor CDM 2015
— prepared before construction work begins, — developed in discussion with, and communicated to, contractors affected by it, — implemented, and — kept up to date as the project progresses; — satisfy themselves that the designers and contractors that they engage are competent and adequately resourced; — ensure that suitable welfare facilities are provided from the start of the construction phase; — take reasonable steps to prevent unauthorised access to the site; — prepare and enforce any necessary site rules; — provide (copies of or access to) relevant parts of the plan and other information to contractors, including the self-employed, in time for them to plan their work; — liaise with the CDM coordinator on design carried out during the construction phase, including design by specialist contractors, and its implications for the plan; — provide the CDM coordinator promptly with any information relevant to the health and safety file; — ensure that all the workers have been provided with suitable health and safety induction, information and training; — ensure that the workforce is consulted about health and safety matters; — display the project notification.	site know the site rules, welfare arrangements, the emergency arrangements and any particular hazards involved in the work. — **Ensure workers are competent** Make sure all workers on your site are trained, experienced and provided with supervisors who are experienced in the type of work to be performed, otherwise you may need to turn them away. — **Ensure plant and equipment is safe** Check that all plant and equipment is being kept in good working order and serviced as recommended. — **Enable communication and cooperation** Talk, listen and take account of the views of workers on site, making changes where they make a good case for improving how work is currently being carried out. — **Ensure first-aid cover** Make sure that you have enough first-aid cover at all times when workers are on site. — **Ensure there are adequate facilities** Make sure that there are enough welfare facilities before the maximum number of workers is on site and that these are maintained and cleaned regularly.

Principal Contractor CDM 2007

What principal contractors don't have to do
Principal contractors don't have to undertake detailed supervision of contractors' work.

Principal Contractor CDM 2015

— **Contribute to the health and safety file**
Pass on important health and safety information to the principal designer for inclusion in the health and safety file. This may include:
  — a brief description of works carried out;
  — any hazardous materials used or retained within the works;
  — information regarding the removal or dismantling of installed plant and equipment;
  — health and safety information relating to maintenance or cleaning of the structure;
  — location of services such as buried cables and gas services;
  — as-built drawings.

Table 4.7 A comparison of contractor duties in CDM 2007 and 2015

Contractor CDM 2007	Contractor CDM 2015
All construction projects — Check that the client is aware of their duties. — Plan, manage and monitor their own work and that of other workers. — Check the competence of all their appointees and workers. — Train their own employees. — Provide information to their workers. — Comply with the specific requirements in Part 4 of the Regulations. — Ensure that there are adequate welfare facilities for their workers. **Additional duties for notifiable projects** — Check that a CDM coordinator and a principal contractor have been appointed and the HSE notified before starting work. — Cooperate with the principal contractor in planning and managing work, including reasonable directions and site rules. — Provide details to the principal contractor of any contractor they engage in connection with carrying out the work. — Provide any information needed for the health and safety file. — Inform the principal contractor of any problems with the plan.	All projects **Where you are the only contractor and no principal contractor is appointed, the contractor must provide to the principal designer:** — the construction phase plan; — evidence of a site induction for all workers; — details of welfare facilities; — arrangements for securing the site and preventing unauthorised access. You must also provide to workers and subcontractors: — information, instruction and training as necessary; — arrangements for consultation and worker engagement; — suitable and sufficient welfare facilities; — clear instructions in the event of serious and imminent danger. **Where there is more than one contractor you must provide to the principal contractor:** — evidence of training and experience relevant to the risks to which the project exposes your workers; — details of arrangements to ensure safe completion of their own works; — requests to subcontract elements of the work; — relevant information for the health and safety file; — evidence that you will provide appropriate supervision which takes into account the level of training, experience and likely behaviour of the workers.

Contractor CDM 2007

— Inform the principal contractor of reportable accidents, diseases and dangerous occurrences.

For all projects, contractors must:
— check clients are aware of their duties;
— satisfy themselves that they and anyone they employ or engage are competent and adequately resourced;
— plan, manage and monitor their own work to make sure that workers under their control are safe from the start of their work on site;
— ensure that any contractor who they appoint or engage to work on the project is informed of the minimum amount of time which will be allowed for them to plan and prepare before starting work on site;
— provide workers under their control (whether employed or self-employed) with any necessary information, including about relevant aspects of other contractors' work, and site induction (where not provided by a principal contractor) which they need to work safely, to report problems or to respond appropriately in an emergency;
— ensure that any design work they do complies with Regulation 11;
— comply with any requirements listed in Schedule 2 and Part 4 of these Regulations that apply to their work;

Contractor CDM 2015

What contractors have to do while working on site
— Take responsibility for dealing with any risks to the safety or health of your workers and others nearby who could be affected by your work.
— Set a personal example by demonstrating good control of your workers, wearing any necessary PPE and selecting safe ways of working.
— You have the right to ask the principal contractor to carry out a site induction for your workers and you. If there is no principal contractor then it is your responsibility to carry out the site induction.
— Provide instructions to your workers on what needs to be done and, importantly, how you intend the work to be done, in which order and with what equipment, especially when it involves working at height. One of the accepted ways of recording this is by carrying out a site briefing with those present signing to show that they have received it.
— If you are not acting as supervisor, or if that role is not being provided by the principal contractor, provide an appropriate supervisor with the necessary training, experience, technical knowledge and leadership qualities to suit the particular work.
— Ensure your workers are briefed, preferably at no greater intervals than at the start of each shift, on what is expected of them, and consider any suggestions from them on better ways of working.

Contractor CDM 2007	Contractor CDM 2015
— cooperate with others and coordinate their work with others working on the project; — ensure the workforce is properly consulted on matters affecting their health and safety; — obtain specialist advice (for example from a structural engineer or occupational hygienist) where necessary when planning high-risk work – for example alterations that could result in structural collapse or work on contaminated land.	— Ensure your workers are complying with the site rules and working in accordance with how you intend the work to be done, accepting any changes required to suit changing circumstances. — Liaise with their principal contractor and keep them informed of any changes to your planned working method in case it has an impact on other plans. — Liaise with other contractors and the principal contractor to cooperate over reasonable suggestions for reducing risks to health and safety on the site. — Carry out checks on your plant and equipment and, where necessary, get it maintained, replaced or changed if, as the job progresses, you think other tools or equipment would reduce the risks to health and safety. — Provide information to the principal contractor about how to safely maintain, isolate, replace or dismantle what you've installed at the end of your time on site.

Contractor CDM 2007	Contractor CDM 2015
	Key messages for contractors — Provide the principal contractor with all requested information. — Work with the principal contractor and other relevant stakeholders to create a positive safety culture. — Eliminate risk at source wherever and whenever possible. — Assess any residual risk and adopt safe systems of work. — Ensure that your workforce has received suitable information, instruction and training. — Comply with the site rules and any other instructions from the principal contractor. — Engage with your workforce and the principal contractor to provide feedback and positive outcomes. — Take responsibility for your own safety. — Consider the effect of actions on the safety of others. — Collect information for the health and safety file as works proceed.

Table 4.8 A comparison of worker duties in CDM 2007 and 2015

Worker CDM 2007	Worker CDM 2015
Involving the workforce in identifying and controlling risks is crucial to reducing the high accident rate associated with construction work. Principal contractors may develop a range of solutions to improve cooperation on site. For example: — involving workers in carrying out site-specific risk assessments; — setting up clearly defined communication channels, to alert the workforce to developments on site; — briefing subcontractors regularly on the work programme; — requiring regular or daily briefings where frontline supervisors brief workers on the work programme and day-to-day risks on site; — setting up formal committees or operatives' forums; — ensuring that issues raised are dealt with and feedback is provided to the workforce. Principal contractors should implement a range of mechanisms to ensure that consultation is effective. Possible ways of doing this include: — engaging with representatives of the workforce (whether appointed by a recognised trade union or elected by the workforce); — establishing health and safety committees or forums; — regular consultation meetings;	What should workers expect while they are on site? Whether you are working on a large civil project or a domestic refurbishment, basic standards and requirements under CDM 2015 must be in place. Some of these are about providing information to workers, some are about engaging with workers on health and safety matters, and some are about providing welfare facilities. You should expect your employer, the contractor or principal contractor to: — provide information about health and safety, including providing a site induction; — ensure that you have the necessary training to work safely and healthily; — consult and engage with workers about site health and safety; — foster a workplace culture of respect and trust – this will result in you and management having a better understanding of any health and safety concerns that are raised; — provide appropriate supervision, depending on work activities being carried out and the skills, knowledge and experience of individuals;

Worker CDM 2007	Worker CDM 2015
— consultation during inductions, daily briefings, toolbox talks, site-wide meetings; — informal methods, for example during site managers' walkabouts, or during senior managers' visits; — procedures to encourage workers to report defects, deterioration in conditions or innovations to raise standards.	— make sure that health hazards are managed as well as safety hazards, and that health risks are clearly communicated to you, along with the details of control measures; — explain the arrangements for cooperation and coordination between workers' employers and other contractors; — make sure that there are adequate, clean and accessible welfare facilities (such as toilets and washrooms) for both men and women; — comply with legal health and safety requirements for construction sites.

Comparing competence in CDM 2007 and 2015

Competence in CDM 2007

Competence and training

This section gives advice about assessing the competence of organisations and individuals engaged or appointed under CDM 2007 – CDM coordinators, designers, principal contractors and contractors.

Assessments should focus on the needs of the particular project and be proportionate to the risks, size and complexity of the work.

To be competent, an organisation or individual must have:

- **sufficient knowledge** of the specific tasks to be undertaken and the risks that the work will entail;
- **sufficient experience** and ability to carry out their duties in relation to the project; to recognise their limitations and take appropriate action in order to prevent harm to those carrying out construction work, or those affected by the work.

Organisations and individuals will need specific knowledge about the tasks they will be expected to perform, and the risks associated with these tasks. This will usually come from **formal or on-the-job training**.

Skills, Knowledge, Training & Experience in CDM 2015

Promoting competence within the construction industry remains a key priority for the HSE, and developing individual competence is crucial to reducing accidents and ill-health. The HSE's vision for competence in the construction industry is one where:

- competence is seen by employers as a long-term issue, building on the basics of selection, training, management of experience and life-long learning. Supervision is vital, but is not a substitute for competence;
- small contractors should only have to complete the minimum amount of paperwork possible to demonstrate their health and safety arrangements at the pre-qualification stage;
- third-party schemes all use the standards for pre-qualification in health and safety set out in Publicly Available Standard 91 (PAS91). Where clients require higher standards this must be explicitly recognised;
- third-party schemes all belong to a common framework of accountability, e.g. the Safety Schemes In Procurement (SSIP) Forum;

Competence in CDM 2007

Appropriate experience is also a vital ingredient of competence. People are more likely to adopt safe working practices if they understand the reasons why they are necessary, and past experience should be a good indicator of the person's/company's track record.

The development of competence is an ongoing process. Individuals will develop their competence through experience on the job and training, part of **life-long learning**. Professionals such as designers, CDM coordinators and advisors should be signed up to a **Continuing Professional Development** programme either through their company or professional institution. This will allow them to remain up to date with changes in legislation and professional practice. Construction tradespeople and labourers should also receive **refresher training or regular training updates** either through an in-house planned programme of learning and development, or a more formal skills-based training programme such as those offered by the CITB in construction skills.

What you must do

All those with duties under CDM 2007 must satisfy themselves that the businesses they engage or appoint are competent. This means making reasonable enquiries to check that the organisation or individual is competent to do the relevant work and can allocate adequate resources to it. Those taken on to do the work must

Skills, Knowledge, Training & Experience in CDM 2015

— clients do not insist, at the pre-qualification stage, on a contractor filling in their own in-house questionnaires, where similar paperwork has already been completed for another client or procurement scheme; clients take seriously their responsibility, at the award stage, to ensure that contractors have the capacity in terms of time, resources and managerial and supervisory capability to deliver the project;
— the site-based workforce is demonstrably qualified through qualifications based on agreed national standards;
— PCs do **not** insist that occasional site visitors, including professionals or ancillary trades, require a competence card.

The HSE plan is to retain a general requirement under the revision of CDM (new Regulation 8) for those appointing others to carry out construction work to ensure that they have received appropriate information, instruction, training and supervision to allow them to work safely. This aligns with the general requirements under Sections 2 and 3 of the Health and Safety at Work Act.

The HSE believes that the competence of construction industry professionals should be overseen by, and be the responsibility of, the relevant professional bodies and institutions, such as the RIBA, ICE, RICS, etc.

Competence in CDM 2007

also be sure that they are competent to carry out the required tasks before agreeing to take on the work.

For notifiable projects, a key duty of the CDM coordinator is to advise clients about the competence of designers and contractors, including the principal contractor that they engage.

Doing an assessment requires you to make a judgement as to whether the organisation or individual has the competence to carry out the work safely. If your judgement is reasonable, taking into account the evidence that has been asked for and provided, you will not be criticised if the organisation you appoint subsequently proves not to have been competent to carry out the work.

How to assess the competence of organisations
Competency assessments of organisations (including principal contractors, contractors, designers and CDM coordinators) should be carried out as a two-stage process.
Stage 1 – An assessment of the company's organisation and arrangements for health and safety to determine whether these are sufficient to enable them to carry out the work safely and without risk to health.

Skills, Knowledge, Training & Experience in CDM 2015

The HSE acknowledges the presentational difficulties associated with removing Regulation 4 and Appendix 4 of CDM, and remains committed to supporting the industry in ensuring its workers are competent.

The proposed removal of the detailed competence requirements is intended to create an environment in which the HSE can work with the industry through non-regulatory approaches to ensure the systems it operates for individual and corporate competence assurance suit the industry's needs.

The proposed approach on corporate and individual competence aligns with work on client capability and procurement and capability of the workforce outlined in *Industrial strategy: government and industry in partnership*.[12]

Competence in CDM 2007	Skills, Knowledge, Training & Experience in CDM 2015
Stage 2 – An assessment of the company's experience and track record to establish that it is capable of doing the work; it recognises its limitations and how these should be overcome and it appreciates the risks from doing the work and how these should be tackled. In order to provide more consistency in the way in which competency assessments of companies are carried out, a set of 'core criteria' have been agreed by industry and the HSE. These are set out in the Appendix. **Stage 1 and Stage 2 assessments should be made against these core criteria**. Organisations that are bidding for work should put together a package of information that shows how their own policy, organisation and arrangements meet these standards. If regularly updated, this information should then be used each time they are asked to demonstrate competence as part of a tender process.	

Table 4.9 Comparing competence in CDM 2007 and 2015

For further information about professional competence and industry support, see Appendix I.

5 Positive behaviours, cultures and approaches

Chapter overview

— Safety Differently – a need for different ways of looking at CDM and safety

— Safety Visually – the visual communication of risk in architecture

Safety Differently – a need for different ways of looking at CDM and safety

While in the past it has been proven that our industry benefits from working within a legislative framework, there is a growing body of thought within the industry and health and safety circles that we have gone as far as we can with reactive systems of safety and paperwork. We need to look at what we do well and how we can enhance and improve this by using innovative approaches and lateral thinking.

There are a number of groups around the world, largely within academia and the construction industry, that are promoting these creative and inspiring 'can-do' approaches to safety. The website Safety, Differently www.safetydifferently.com was set up by Daniel Hummerdal based on the work of Sidney Dekker, Professor at Griffith University in Australia and thought leader on safety. It aims to encourage sharing of innovative and critical approaches to safety thinking in all industries, but has sparked some ground-breaking work in the construction industry in particular.

Safety-I and Safety-II

One particularly innovative approach to safety is that promoted by the academic Erik Hollnagel of the University of South Denmark, who develops the concept of Safety, Differently into wholly intelligible Safety-I and Safety-II approaches, both of which have an application to architecture as defined below:

Safety-I

Safety has traditionally been defined as a condition where the number of adverse outcomes was as low as possible (Safety-I). From a Safety-I perspective, the purpose of safety management is to make sure that the number of accidents and incidents is kept as low as possible, or as low as is reasonably practicable. This means that safety management must start from the manifestations of the absence of safety and that, paradoxically, safety is measured by counting the number of cases where it fails rather than by the number of cases where it succeeds. This unavoidably leads to a reactive approach based on responding to what goes wrong or what is identified as a risk or as something that could go wrong.

Safety-II

Focusing on what goes right, rather than on what goes wrong, changes the definition of safety from 'avoiding that something goes wrong' to 'ensuring that everything goes right'. More precisely, Safety-II is the ability to succeed under varying conditions, so that the number of intended and acceptable outcomes is as high as possible. From a Safety-II perspective, the purpose of safety management is to ensure that as much as possible goes right, in the sense that everyday work achieves its objectives. This means that safety is managed by what it achieves (successes, things that go right), and that likewise it is measured by counting the number of cases where things go right. In order to do this, safety management cannot only be reactive, it must also be proactive. But it must be proactive with regard to how actions succeed, to everyday acceptable performance, rather than with regard to how they can fail, as traditional risk analysis does.[13]

Architects usually design by an eclectic process, and from precedent rather than from first principles. This is a different process from the structural or civil engineer, who needs to go back to the mathematical calculations to prove that a certain structural design will or will not stand up.

The use of vernacular design handed down through centuries of worldwide civilisation, together with the ground-breaking technology of modern products, systems and methods, are the essential ingredients for any new design. Thus very rarely is the construction process so innovative that it cannot be managed in a safe way.

Safety-II is a creative process in which these safety essentials are identified to achieve a given design, working with the team and contractor to ensure that it is buildable and maintainable.

Safety-I tends to be a reductive process by which features, elements or materials are regarded as being potentially injurious and so are eliminated from a design.

The intention is not to completely disregard Safety-I or introduce Safety-II everywhere but to introduce a proportionate balance between the two safety approaches to achieve a tolerable level of safety on any project.

Safety Visually – the visual communication of risk in architecture

In order to harness the power of Safety-II the architectural profession is examining the ways that CDM issues can be analysed at design stages on all projects in an efficient, collaborative and creative manner. In this way acceptable CDM design outputs should be produced for implementation during construction and maintenance stages.

One of these methods, entitled Safety Visually, has been developed by Scott Brownrigg and won peer review accreditation at the APS CDM Innovation Awards 2013. The principle is based on previous industry-wide and legislative aspirations to review, record and discuss CDM issues in the context of the entire design project, rather than just in relation to health and safety issues.

This process has evolved into CDM Visually, a methodology that identifies significant risks at design stages in a visual way, and communicates risk jargon-free to all stakeholders. Its application is demonstrated by examples of live projects of various sizes being undertaken at the time of writing. This methodology creates a powerful opportunity to change the management of risk and the integration of safety systems in all architectural projects. It has resonated especially with the key design and construction stakeholders, facilitates team collaboration by predominantly visual communication techniques and meets statutory and operational expectations.

Identifying hazards and risks

The concept of identifying risk visually is commonplace on roads, buildings, stations, maps and even construction sites, so why not on construction drawings and associated documents? This is not just to assist contractors, but also the design and project team during design development (see Fig. 5.1).

Industry context

The Safety Visually principles are already being used in parts of the industry as an alternative method of identifying hazards and associated risks, minimising unacceptable risks and passing on possible risk control measures to contractors for managing during the construction phase, with a minimum of narrative documentation (see Fig. 5.2). This reflects the need to recognise 'acceptable or tolerable risks' within projects and the construction industry, and to help designers to go only 'so far as is reasonably practicable' (SFARP) to mitigate these significant issues as a collaborative team. This visual analysis process allows these often difficult-to-agree concepts and risk pathways to be

Figure 5.1 Safety communication methods

Use to warn of significant design risks and site hazards

 Use to avoid or prevent a particular action

 Use to encourage a particular action

 Use to convey some relevant CDM information

Figure 5.2. These icons provide four simple visual aids when annotating plans and drawings

discussed collectively and recorded in a manner that everyone can easily understand, including operatives and foreign workers. This process should help in bringing together a widely divergent part of our industry. In turn this should create more cohesive team-working expectations and better health and safety outcomes, and facilitate the construction of more innovative and creative designs, previously challenged by onerous health and safety expectations.

The method
Simply put, Safety Visually identifies significant risks at each design stage in a visual way by brainstorming 3D images and drawings (see Figs 5.3 & 5.5). It communicates risk to all stakeholders clearly and facilitates open discussion of design risk resolution options. It also allows access to more detailed information (if required). It communicates risk reduction methods to others and feeds back to the industry. By facilitating collaborative working, Safety Visually has evolved as an easy-to-comprehend project storyboard (see Figs 5.8–5.19). This helps all project participants to engage in a blame-free, open and transparent dialogue regarding significant CDM issues throughout project preparation stages in design team meetings. The developing visual design stage drawings or options matrix directs the tendering contractors towards transient or hidden constructability issues that would not necessarily be easy to glean from the architectural drawings, or from the written pre-construction information. It provides a financially level playing field

for all tenderers and creates a benchmark of CDM expectations for the construction phase. Contractors also develop their project execution stage information, identifying their preferred construction and risk management methods in a similarly visual way as feedback to the client and design team, which can also later be used by the workforce – who arguably need it most.

Practical application

The CDM Visually process is being used on a wide variety of projects and is continually being refined using feedback from internal and external sources. On the Southampton Solent University and North Hertfordshire College projects, the process has identified a number of significant risks relating to the existing site, proposed building design and the client's safety expectations of the completed building, particularly relating to future maintenance (see Figs 5.3–5.19).

CDM Options Analysis Matrix (Fig. 5.8)

This records information in a logical format for wide discussion by the construction team, client team and other stakeholders, with recommended solutions and options considered to confirm a SFARP approach. As the project progresses and as an increasing level of detail evolves, these options are rationalised into one safe but aesthetically appropriate solution for the project. This document works in conjunction with the identification of the significant issues on the project surveys, site plans, architectural drawings and visualisations and forms a progressively developed 'safety specification' of significant issues. It is important **not** to identify risks that are purely related to a normal trade contractor or main contractor construction issues that are within the capability and training parameters of experienced contractors.

Alongside its application across a variety of projects, case studies are being produced to demonstrate lessons learned and good practice for dissemination internally, as well as externally to organisations such as DIOHAS and WREN (see Figs 6.1–6.5 in the next chapter).

The following visualisations, drawings and coordinated safety storyboards illustrate the Safety Visually process.

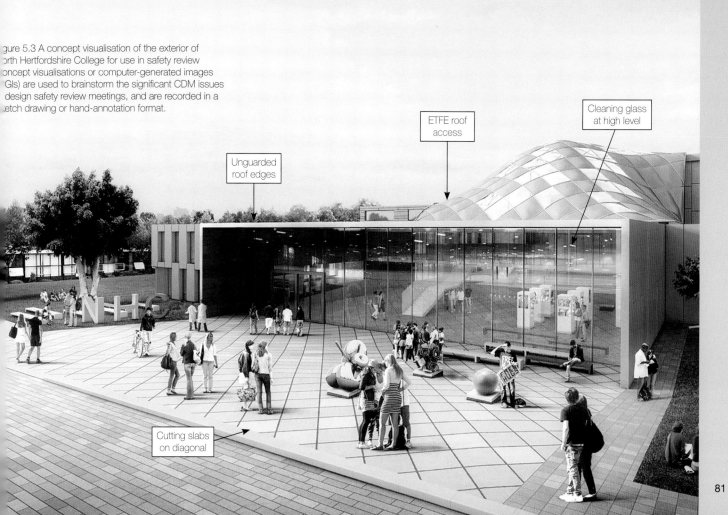

Figure 5.3 A concept visualisation of the exterior of North Hertfordshire College for use in safety review. Concept visualisations or computer-generated images (CGIs) are used to brainstorm the significant CDM issues in design safety review meetings, and are recorded in a sketch drawing or hand-annotation format.

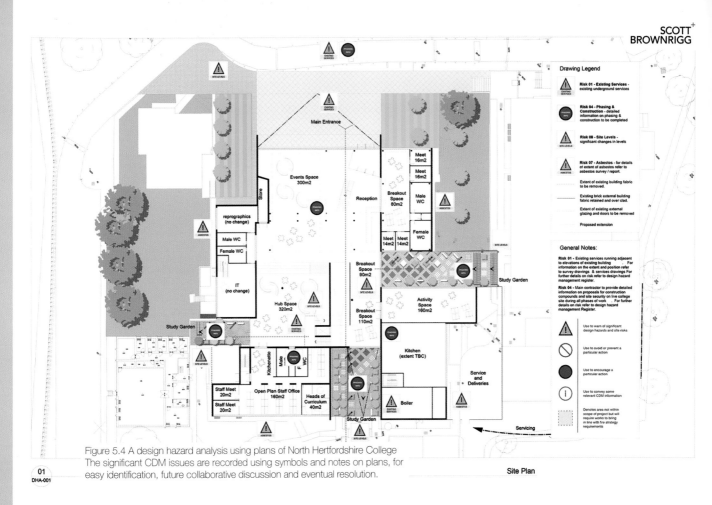

Figure 5.4 A design hazard analysis using plans of North Hertfordshire College
The significant CDM issues are recorded using symbols and notes on plans, for easy identification, future collaborative discussion and eventual resolution.

Figure 5.5 An interior visualisation of North Hertfordshire College for use in safety review. Concept interior visualisations are used to identify significant CDM issues in relation to buildability, maintainability and usability in the same way as the externals.

Figure 5.6 Design hazard analysis using elevations of North Hertfordshire College
Sections and elevations are annotated and symbols added for identification. Appropriate methods of access for maintenance or construction are added.

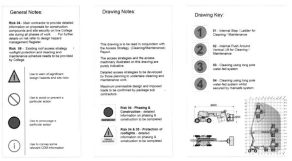

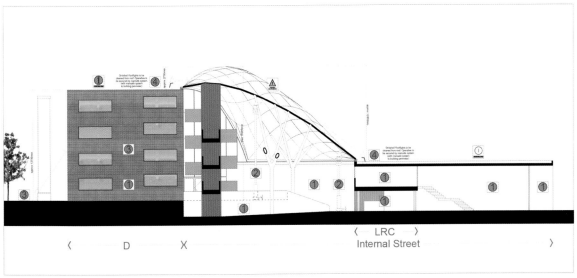

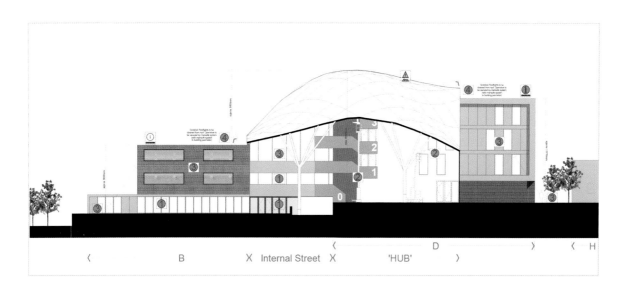

Figure 5.7 Southampton Solent University lecture theatre pod in the atrium
Unusual, innovative or special design features are identified for particular CDM consideration in a visual format for more detailed collaborative design discussion and CDM resolution.

Designer CDM Analysis and Options Matrix – Hazard Identification and Significant Risk Management

Project & No:	Solent University Campus		Work Stage:		Revision & Date:	16		17 November 2014
HAZARDS and SIGNIFICANT RISKS	BUILDING FORM, MATERIAL, ACTIVITY, LOCATION	ELIMINATE or AVOID risks (During early design stages) SFARP	REDUCE or MINIMISE risks ALARP by: (During all design stages) Safe systems of work & protection		INFORMATION To be provided with the design e.g. specialist design & client input	CONTROL METHODS Contractor or client management systems		ACTIONS & DATES Other specialist guidance & comments

Agreed | Agreed in principle | Not yet agreed

2.3.3 LEVEL 1

Annotations on plan:
- Suggested construction stage material, vehicle and personnel route into atrium from site compound. Levels of terracing to be considered for actual location and viability
- Millais scaffolding zone
- Main Millais entrance and fire exit during construction
- Fire exits closed during construction
- FLOOR G
- External Area

CDM OVERVIEW ISSUES

- Integration with existing deck and flat roof with fragile roof lights
- Existing services
- Adjacent foundations and structural interfaces
- Construction access into atrium zone?
- Levels, sloping site
- Works inside, on top of and elevation of Millais during occupation?
- **Extent of decant of Millais during construction of the block etc. client PC**
- Construction phase fire strategy PC
- Diversion of Millais access and escape routes C/PC
- **Asbestos survey of Millais – how much impact on design?**
- Temporary protection of Millais elevation
- Roof works to Millais with waterproofing and plant implications
- Temporary structural works
- Access doors into atrium for roof access MEWPS and large pieces of equipment for displays, exhibitions, easily removable panels to front elevation

Team sign-off status	Client		Architect		Struct. Eng		Services Eng		P. Contractor
Others	PM		CDM–C		Landscape		Cost Consultant		

Figure 5.8 Designer CDM analysis and options matrix for Southampton Solent

If there is too much detailed CDM information to capture on the project drawings a CDM analysis options set of visual documents can be prepared using a logical template format and annotation of project drawings. The following documents give you an example of what is possible.

Designer CDM Analysis and Options Matrix – Hazard Identification and Significant Risk Management

Project & No:	Solent University Campus			Work Stage:			Revision & Date:	16		17 November 2014
HAZARDS and SIGNIFICANT RISKS	BUILDING FORM, MATERIAL, ACTIVITY, LOCATION		ELIMINATE or AVOID risks (During early design stages) SFARP	REDUCE or MINIMISE risks ALARP by: (During all design stages) Safe systems of work & protection			INFORMATION To be provided with the design e.g. specialist design & client input		CONTROL METHODS Contractor or client management systems	ACTIONS & DATES Other specialist guidance & comments
										Agreed / Agreed in principle / Not yet agreed
2.4 3D view of pod and atrium				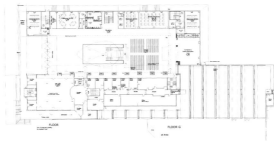						**CDM OVERVIEW ISSUES** High-level access issues for cleaning and maintenance
										Maintenance access to elevations
										Maintenance to pod
										Balcony pods and fronts – off-site manufacture to be considered for speed and safety
				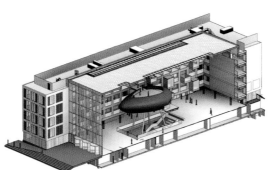						Atrium lighting access not to be within roof but in accessible side wall or building roof locations
										Walkway at high level adjacent to atrium for roof vent access and maintenance externally
										Fire stairs available for construction stage escape and existing Millais stairs for existing population evacuation
										Prefabrication of stair flights and cranage into location. Temporary or permanent handrail protection to be included
Team sign–off status	Client			Architect			Struct. Eng		Services Eng	P. Contractor
Others	PM			CDM–C			Landscape		Cost Consultant	

Figure 5.9 Designer CDM analysis and options matrix for Southampton Solent

3D drawings with cutaways and part plans are very useful to understand the context of special CDM issues for team collaborative discussion on buildability, maintainability and usability.

Designer CDM Analysis and Options Matrix – Hazard Identification and Significant Risk Management

Project & No:	Solent University Campus		Work Stage:		Revision & Date:	16		17 November 2014
HAZARDS and SIGNIFICANT RISKS	BUILDING FORM, MATERIAL, ACTIVITY, LOCATION	ELIMINATE or AVOID risks (During early design stages) SFARP	REDUCE or MINIMISE risks ALARP by: (During all design stages) Safe systems of work & protection		INFORMATION To be provided with the design e.g. specialist design & client input	CONTROL METHODS Contractor or client management systems		ACTIONS & DATES Other specialist guidance & comments
						Agreed	Agreed in principle	Not yet agreed
2.3.3 LEVEL 1						Primary structural assembly inside atrium with tower crane		
						Crash deck below pod to construct underside		
						Cladding assembly on site with spider crane		
						Lifting and removing spider crane or hoisting it in parts		
						Temporary propping of upper slab to support spider crane		
						Jointing of cladding panels as off-site finish		
						No confined space working in underbelly		
						Prevention of 'drinks tins' sliding off the pod		
						Bridge assemblies and high-level access to services		
						Glass balustrades to bridges need temporary protection		
						Cladding finish to be completed off site, damage repairs on site only from MEWP		
						Future cleaning from MEWP		

Team sign-off status	Client		Architect		Struct. Eng		Services Eng		P. Contractor
Others	PM		CDM–C		Landscape		Cost Consultant		

Figure 5.10 Designer CDM analysis and options matrix for Southampton Solent

Additional images and sketches can be added to the matrix to further clarify the design intent and CDM issues as demonstrated above

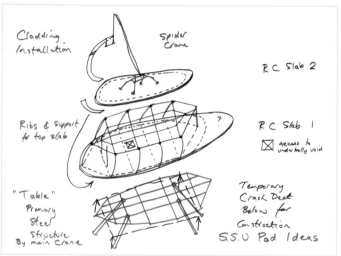
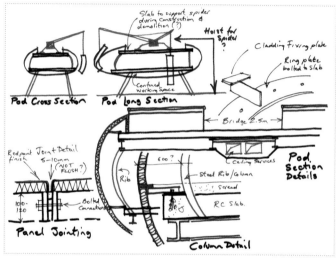

Figures 5.11–5.14 During design reviews or design team meetings sketches are produced on flipcharts, whiteboards or large drawings for collaborative discussion and contributions. These are scanned or recorded photographically and used in the matrix or issued with meeting notes or minutes

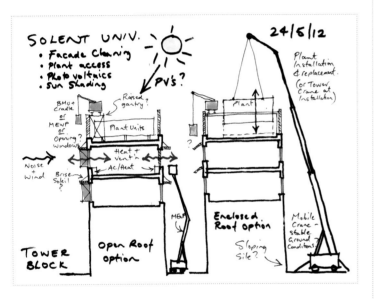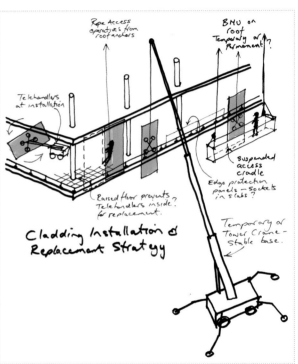

Positive behaviours, cultures and approaches

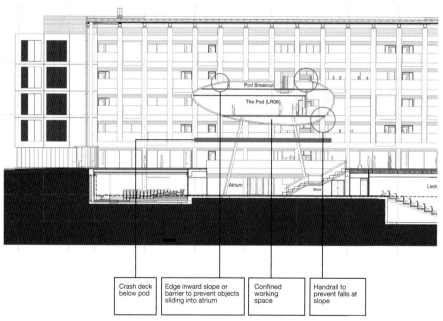

Figure 5.15 Precedent research of other similar designs can be used for comparative purposes and the CDM design evolution can take place in an informed context. Detailed annotation of drawings can highlight important issues

Designer CDM Analysis and Options Matrix – Hazard Identification and Significant Risk Management

Project & No:	Solent University Campus	Work Stage:		Revision & Date:	7	16 August 2014
HAZARDS and SIGNIFICANT RISKS	BUILDING FORM, MATERIAL, ACTIVITY, LOCATION	ELIMINATE or AVOID risks (During early design stages) SFARP	REDUCE or MINIMISE risks ALARP by: (During all design stages) Safe systems of work & protection	INFORMATION To be provided with the design e.g. specialist design & client input	CONTROL METHODS Contractor or client management systems	OTHER SPECIALIST GUIDANCE & COMMENTS e.g. ACTIONS & DATES
						Agreed / In principle / Not yet agreed
1.0.2 PHASE 1 Construction Site Set-Up Plan • Vehicular access • Car parking • Welfare • Site storage • Crane locations • Site roads • Boundary fencing	*[site plan showing Cars, Cars, Site compound, The Site, Service Yard, Car]*		Impossible to eliminate all hazards or assess risks until full site information available All phases of project to be considered but Phase one to be considered in detail Site set-up workshop required to analyse and capture the issues and find most cost-effective but safe solution Site waste management strategy arrangements to be considered. Locations of skips for delivery, ease of use and collection Temporary parking?	On-site traffic management solutions are a major requirement for this project at all phases The proposed two-way system for the final 'in-use' traffic circulation is NOT recommended good practice for site traffic phases Car parking for site staff and academic usage needs to be managed. Major themes: • Keeping pedestrians and vehicles apart • Minimising vehicle movements • People on site • Turning vehicles • Max. visibility • Signs and instructions	Reference: HSG144 *The safe use of vehicles on construction sites*[14] Site layout for all vehicles & pedestrians set out by projects team in principle at tender Contractor to make proposals for site set-up and phasing prior to commencement Client project team and developer contractor to agree a final site set-up plan to include the issues raised
Team sign-off status	Client	Architect		Struct. Eng	Services Eng	P. Contractor
Others	PM	CDM-C		Landscape	Cost Consultant	

Figure 5.16 Client design site set-up plan
The key strategic CDM information required for any project is the site set-up plan, which sets the overall parameters for the contractor to establish the logistics of construction access, welfare and storage. The early design decisions of layout and location of the final building can greatly influence the cost and efficiency of these logistics.

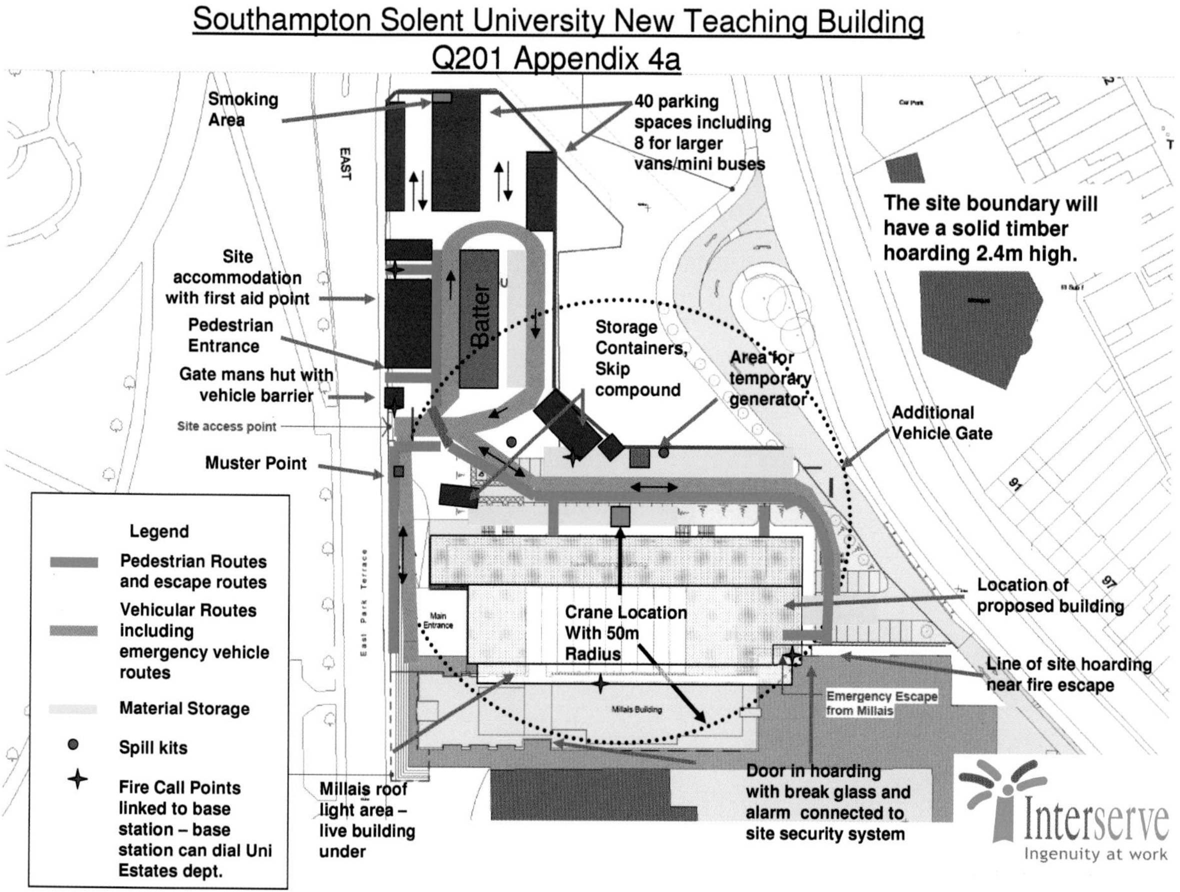

Figure 5.17 Contractor's site set-up plan
The contractor can take the client and principal designers in-principle site-set up proposals and develop a more detailed site establishment plan identifying any issues of concern and mutual benefit. This is especially significant in extension projects or where existing activities or buildings have a close proximity to the works.

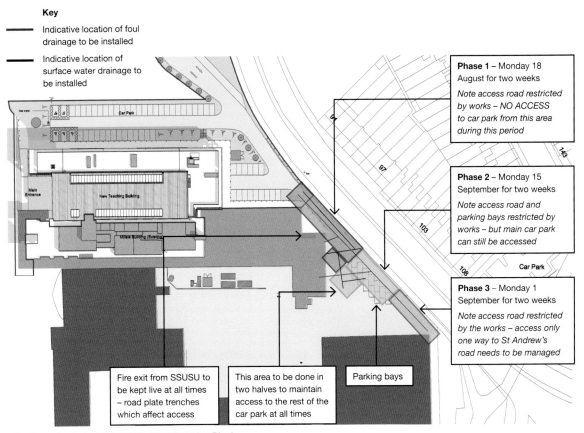

Figure 5.18 Contractor's detailed phasing plans
The details of specific mini-projects, timescales and handovers within the context of the overall project can be identified on larger-scale drawings with suitable annotation. These can be collaboratively produced and commented on by others after being issued. These help the client to understand the intended activities, and how these details may affect the design proposals.

Designer CDM Analysis and Options Matrix – Hazard Identification and Significant Risk Management

Project & No:			Work Stage:		Revision & Date:			
Beaufort House, Guildford							Tender	9 March 2013
HAZARDS and SIGNIFICANT RISKS	BUILDING FORM, MATERIAL, ACTIVITY, LOCATION	ELIMINATE or AVOID risks (During early design stages) SFARP	REDUCE or MINIMISE risks ALARP by: (During all design stages) Safe systems of work & protection		INFORMATION To be provided with the design e.g. specialist design & client input	CONTROL METHODS Contractor or client management systems	ACTIONS & DATES Other specialist guidance & comments	
							Agreed	Agreed in principle	Not yet agreed

1.0 Scope of Works

- Removal of existing structural brickwork external walls to courtyard
- Three-storey extension to existing building to infill courtyard (extent defined by the black columns on the plans) above an existing basement
- Redesigned core relocated to the centre of the building
- Improved reception area – including remodelling of existing arched entrance and adding full-height glazing strip
- Replacing two second-floor rear porthole windows, adding curtain wall glazing
- Revised roof plant room strategy
- 'Thinning' landscaping to front and side elevations and revised soft landscaping scheme to rear area
- Internal office fit-out to Cat. A including chilled beams, plasterboard ceilings and raised floor
- Complete new A/C and M&E services except for basement smoke and fumes ventilation
- Removing screed to existing office floors for maximisation of head height
- Making good mansard roof and hard landscaping where disturbed by works
- New plastered finishes to existing escape stair cores over fairfaced brickwork. Services to be surface mounted to avoid chasing
- Concrete soffit repair to ground floor. Structure slap soffit in basement where damaged, plus painting of finished surface
- Miscellaneous structural modifications to basement plant rooms, structure and slabs
- Existing bridge link to adjacent building to be retained

Team sign-off status	Client	Architect	Struct. Eng	Services Eng	P. Contractor
Others	PM	CDM–C	Landscape	Cost Consultant	

Figure 5.19 The same process can be used on smaller-scale and traditional-build projects, as demonstrated above, to capture the CDM issues in one document for ease of discussion and clarity

6 Hazard awareness and risk identification

Chapter overview

— Hazard and risk identification tools
— CDM 2015 and the use of red, amber and green lists
— Risk and hazard logs that are fit for purpose
— Safety design reviews
— Avoiding and minimising rather than eliminating and reducing risks

In this chapter we assess how current target-setting legislation for 'risk reduction', combined with the huge amount of information on risk, can be overwhelming for designers when trying to assess risk on their projects.
      Below we look at several approaches to assessing risk and hazards on projects that will make this process more effective, as well as providing suitable resources. They are designed to be straightforward and clear in their intentions, encouraging discussion, collaboration and creative solutions.

Target-setting legislation
The Health and Safety at Work Act 1974 introduced the concept of target-setting legislation with risk reduction – in preference to prescriptively identifying individual issues that cause harm and considering how they could be avoided. Although the intentions were good, the introduction of the idea of risk assessment has probably created as much confusion and over-zealous practice as it has actually improved the control of risk. In an architectural design context the intention was always to assess the H&S risks associated with any task in relation to the qualitative complexity and aspirations of the design concept intent (i.e. the other project influencing factors). The idea was not to assess the quantitative terms of its severity, likelihood and consequence in relation to other risks, as this is a construction stage approach. In essence we need to weigh the tolerability or acceptability of the design risk with the effort and resources needed to manage the risk. However, the risk-averse sections of the construction industry have interpreted this assessment process as a quantitative numerical exercise. They see that identifying high-risk issues, activities or materials and eliminating them or replacing them with lower-risk activities, elements or materials will reduce the number of accidents. This exemplifies the disconnect between

aesthetically trained designers and those with more numerate engineering, health and safety or construction training.

Information overload
The intentions of target-setting legislation presuppose that those undertaking the risk assessment are fully aware of the extent of legislation and guidance that relates to each individual risk they encounter. In the 40 years since 1974 the level of safety expectation has risen progressively as new 'risk issues' are researched and guidance published. The sheer volume of this material in a wide number of reports and documents, often owned by independent organisations and some only available at a cost, has led to a mixture of information overload and inaccessibility. This creates a challenge for busy designers to find, understand and integrate the recommendations and information into projects within practicable timescales. By comparison the Building Regulations,[15] the origins of which date back to the 1666 Great Fire of London, have a target-setting objective but provide prescriptive advice in the Approved Documents.[16] Designers can use their experience and imagination to meet the intentions of the Building Regulations in a variety of ways, providing an equivalency can be proven. The CDM Regulations however are solely target-setting with a few general examples, which although helpful do not provide a fully comprehensive database of all the issues with which a design team needs to engage.

Hazard and risk identification tools

Proportionate and practicable information and responses
Developed in collaboration within the architectural design fraternity at the RIBA and DIOHAS, the following templates are provided to help identify design-relevant 'significant issues'. These lists are not comprehensive but give an indication of the sort of issues and information to identify on projects, and help you develop a strategy to record and deal with them in a straightforward manner.

Below is an example of a Design stage hazard awareness and risk identification checklist, with some examples of potential risks and how to deal with them, which can be used by designers to highlight risks on a particular project according to specific categories (e.g. fire) and levels of risk (red, amber and green, or 'RAG') along with notes for action. This section is the document identification and control header.

Design Stage Hazard-Awareness and Risk-Identification Checklist (HARI)			
Project Number		Key to status of design	RAG
Location		Risk tolerable	RM/RR
Building Type		Risk to be further reviewed or agreed	
Date of Review		Risk not tolerable	
Date of Scheme/Detailed Design Review		Risk review pending or N/A	

RM – Risk mitigated indicates that the risks have been assessed and are judged to be no more significant or unusual than a capable contractor would be expected to manage or to be aware of, or that mitigation measures have been included in the specification and drawings.
RR – Residual risk has been identified and considered tolerable in relation to the concept design intent that cannot be designed out.

Table 6.1 Design stage hazard awareness and risk identification checklist

The project information below can be collated into the following register for simple communication with contractors and others.

Hazard Awareness and Risk Identification Checklist (HARI)				
Risk		**Designers to identify and consider the following significant risks and other factors – SFARP**	**Comments, Actions, References**	**RAG Colour**
A		**Catastrophic risks – site specific and generic**		
	Structural collapse	– Existing and adjacent buildings, structures, party walls, retaining walls, etc. during construction, and permanent condition – Underground voids such as tunnels, vaults, mines, old workings, wells, etc. – Existing fabric of buildings during refurbishments under temporary loadings of scaffold, materials or plant – Construction cranes, scaffolding or trees in poor condition falling in high winds – Temporary works, including all types of scaffolding, shoring, propping. Ensure temporary works designer and coordinator appointed	*RR834* – Preventing catastrophic risks in construction[17] *CIRIA Publication C699* – Guidance on catastrophic events in construction[18] *BS 5975:2008 – Code of practice for temporary works procedures and the permissible stress design of falsework*[19]	
	Fire	– Fire during construction – means of escape for operatives (e.g. stairs, routes) and existing occupants (if in use). Access for fire-fighting appliances and personnel – Temporary fire protection to prevent fire spread to adjacent properties in construction phase, e.g. timber frame protection and unprotected area calculations	*HSG168 – Fire safety in construction*[20] *UKTFA (STA) guidance on fire spread*[21]	

Hazard Awareness and Risk Identification Checklist (HARI)

Risk	Designers to identify and consider the following significant risks and other factors – SFARP	Comments, Actions, References	RAG Colour
Water	– Rivers, canals, culverts, storm drains on site or adjacent – Temporary/permanent railing as an early activity – Contractor to issue a safe system of work after seeking further advice, including consideration of lifejackets, throw-lines, torpedo floats and grab rings, floatation devices, rescue and safety harnesses, rescue boats, communications to raise alarm – This should ensure that there is an adequate method of preventing falls on to the mud or into the river and for retrieving persons – Identify flooding risk – occupier may need to join the Environment Agency Flood Warning System and establish an evacuation procedure – Stagnant water from rivers, canals, culverts, drains. Risk of contamination to watercourses by spills on site. Infection/Weil's disease	*Reference to HSE prosecution of the Environment Agency, following a fatal accident on the banks of the River Witham on 12 September 2001[22]*	
Wind or extreme weather	– Identify likelihood of extreme weather conditions on site in long and short term which may influence the design and construction methodology		
B	**Significant risks – site specific**		
Noise, vibration and disturbance to neighbours	To occupied buildings and housing nearby, hospitals, churches, cemeteries, schools, care homes, etc. – Piling operations – hammered/driven or augered? Noise disturbance or damage to surrounding buildings – Risk of disturbing those in vulnerable sectors of the community – Limits on hours of site work		

Hazard awareness and risk identification

Hazard Awareness and Risk Identification Checklist (HARI)

Risk	Designers to identify and consider the following significant risks and other factors – SFARP	Comments, Actions, References	RAG Colour
Health and respiratory injuries from materials and dust	To local environment, all neighbours, operatives, and site personnel — Identify asbestos materials and other contaminants in existing buildings, e.g. horsehair plaster — Ensure adequate surveys and appropriate contractors used to identify dangerous materials on site and/or their removal — Avoid or minimise chasing, cutting blocks and masonry etc., unless procedures in place for dust suppression or vacuum extraction, or enclosure and extraction — Demolitions and concrete breaking – minimise works and damping-down processes — Avoidance of screed removals where possible — Lead paint in existing buildings – avoidance of dry sanding, drilling or cutting. Surveys may be necessary	*L143* – Managing and working with asbestos[23] *Control of Asbestos Regulations 2012*[24]	
Site access and construction facilities	— Surrounding roads and access roads to be considered for safe access and egress of staff, materials and waste. Low or weak bridges, narrow roads, overhanging trees or cables, etc. — Welfare facilities to be considered in terms of location, services, and convenience of contractors and existing users. Phasing of facilities as project progresses to be considered — Materials unloading and waste storage areas to be considered — Static and mobile crane sizes, locations, weights and access to be considered in principle, away from possible vehicle impacts, and collapse on occupied premises where possible — Consider location, convenience and security of construction team vehicular parking, materials and tools storage facilities	*HSG150 – Health and safety in construction*[25]	

Hazard Awareness and Risk Identification Checklist (HARI)

Risk	Designers to identify and consider the following significant risks and other factors – SFARP	Comments, Actions, References	RAG Colour
Project phasing	— Location of existing buildings, car parks and roads to be considered in relation to temporary and future site arrangements in terms of cost, convenience and safety — Transient delivery or collection locations and times, such as school runs or food deliveries during construction period		
C	**Significant risks – site generic**		
Electrical interference to radio equipment	— Interference with neighbouring hospitals, ambulance stations, airport — Specify no site radios where safety is critical	*This is subject to site-specific risk assessment by PC*	
Injury to trespassers	— Hoarding to/fencing-off of site and deep excavations — Security measures to scaffolding, detectors, CCTV		
Animals, vegetation, poisonous bites	— Consider risk of adders, hornets and wasps' nests from adjacent sites — Aggressive seagulls or other birds, particularly during nesting season — Consider bats, birds, newts and other fauna with regard to their breeding seasons and relocations — Consider poisonous or aggressive vegetation such as Japanese knotweed that can take considerable time to eradicate, or cause significant damage if ignored — Consider tree roots and avoidance of killing trees during construction and causing structural damage to permanent structures	*These environmental safety issues are included in BREEAM requirements[26] but are part of an ethical and criminal requirement to protect endangered and other species from harm. Some can also harm operatives and users. Identification is the key*	

Hazard awareness and risk identification

Hazard Awareness and Risk Identification Checklist (HARI)

Risk	Designers to identify and consider the following significant risks and other factors – SFARP	Comments, Actions, References	RAG Colour
Injury to pedestrians	— Rights of way adjacent to or across site. Temporary closures — Adjacent railways — Pavements – protecting the public and materials		
Electric shock	— Cables, street lighting, even if assumed disused	*BSI PAS 128 Specification for underground utility detection, verification and location*[27]	
Falls on slopes	— Identify steep falls, slopes and embankments which can increase site access costs and difficulties during construction and maintenance		
D	**Existing services and utilities**		
Stomach infection	— Water supplies potable? Water quality tested?		
Fire	— Water supply/hydrants available for construction fire use		
Electric shock	— Overhead cables – barriers specified — Underground cables – investigation to confirm position as survey drawing and utility company information. Identify possibility of unknown live services. Further surveys and hand digging likely to be required — Cables assumed disused, to be identified and checked	*BSI PAS 128 Specification for underground utility detection, verification and location*	

Hazard Awareness and Risk Identification Checklist (HARI)

Risk	Designers to identify and consider the following significant risks and other factors – SFARP	Comments, Actions, References	RAG Colour
Explosion	– Gas pipes – investigation to confirm position as survey drawing and utility company information. Identify possibility of unknown live services. Further surveys and hand digging – Propane cylinders	*BSI PAS 128 Specification for underground utility detection, verification and location*	
Uncertainty	– Surveys of utilities are not totally reliable. On-site checks to be made	*BSI PAS 128 Specification for underground utility detection, verification and location*	
Excavation	– Encourage use of vacuum excavation equipment in proximity of dangerous or multiple utilities locations		
E	**Contamination and buried objects**		
Contamination	– Buried tanks/petrol interceptors – Purging or chemical cleaning of existing services		
Explosion	– Buried ordnance – check local authority records and bomb maps, or obtain a survey. Site detection/probing methods may be necessary prior to excavations – Methane and other ground gases risk		
Musculo-skeletal injuries	– Identify particularly heavy objects or ones that are difficult to handle, remove or break up		

Hazard Awareness and Risk Identification Checklist (HARI)

Risk	Designers to identify and consider the following significant risks and other factors – SFARP	Comments, Actions, References	RAG Colour
F	**Working space/working platforms**		
Falls from height	— Design to allow early installation of permanent floors, edge protection and guarding to holes and penetrations. Fixing holes for edge protection — Allow for work to be carried out at ground level or from permanent floor levels where possible		
Manual handling/ musculo-skeletal injuries	— Avoid site layouts or dictating construction methods that limit space for access where reasonably practicable — Consider type of access platforms, scaffolds and mechanical lifting aids that are appropriate for the design to be constructed and maintained		
Fire/ emergency evacuation	— Design for early installation of stairs and fire compartments where possible. Prefabrication can assist this process		
G	**Confined spaces**		
Entry into confined spaces	— Silos, sewers, ductwork, unventilated rooms, storage tanks, basements, etc. — Avoid creating confined spaces where possible — Minimise operations involving hazards in confined spaces, e.g. welding, cutting, etc. — Minimise work in basements and the need for deep trenches	*If confined spaces are essential avoid the need for access or consider safe methods of use*	

Hazard Awareness and Risk Identification Checklist (HARI)

Risk	Designers to identify and consider the following significant risks and other factors – SFARP	Comments, Actions, References	RAG Colour
Lack of oxygen/ poisonous gases (including from residues left)	— Avoid creating, and minimise operations involving, hazards in confined spaces where possible — Avoid entry to confined spaces where possible. If entry is unavoidable, follow a safe system of work and put in place adequate emergency arrangements before work starts — Provision of ventilation and testing the air — Provision of breathing apparatus — Permit to work scheme	*L101 Safe work in confined spaces*[28] *Confined Spaces Regulations 2007*[29]	
Drowning	Due to liquids and solids that can suddenly fill the space — Rescue harnesses and lifelines — Communications to raise alarm		
Fire and explosion	— Check size of emergency exits — Non-sparking tools and 25V max. equipment		
Noxious fumes	— Avoid specifying applied finishes in confined spaces where possible — Avoid specifying hazardous materials/substances for application in confined spaces		
H	**Erecting structures**		
	Steelwork, in-situ reinforced concrete, pre-cast reinforced concrete, pre-stressed concrete, timber, masonry, brickwork, blockwork, roof structures, stairs		

Hazard Awareness and Risk Identification Checklist (HARI)

Risk	Designers to identify and consider the following significant risks and other factors – SFARP	Comments, Actions, References	RAG Colour
Collapse/ temporary instability	– Avoid designs which involve temporary instability during construction or specify erection sequence including details of temporary support measures required – Temporary props and bracing – high winds		
Falls from height	– Maximise pre-fabrication, pre-casting, use of simple intrinsically safe connection details, and allow for early installation of floors, roof decks, stairs, edge protection, etc. to minimise risk from high-level working – Detail to allow easy connection of safety lines, harnesses etc. – Specify easily achievable tolerances where possible	BS 8560 Code of practice for the design of buildings incorporating safe work at height[30]	
Collapse – construction loadings	– Identify construction loadings on drawings together with any temporary support requirements, e.g. high walls, unbraced structures – Provide adequate tender information		
Manual handling/ musculo-skeletal injuries	– Heavy blocks/stone sections/windows components/lintels and window cills – lifting hooks/component size – under 20kg		
Handling major components	– Consider access, storage, erection procedures and lifting details for large or awkwardly shaped components – Consider composite roof panels for easier handling in high winds		
Falling materials	– Design temporary works to avoid falling materials – Tethering of tools to be recommended		

Hazard Awareness and Risk Identification Checklist (HARI)

Risk	Designers to identify and consider the following significant risks and other factors – SFARP	Comments, Actions, References	RAG Colour
J	**Materials/substances/components generally**		
Manual handling/ musculo-skeletal injuries	— Design to allow mechanical handling where possible (avoid blocks weighing over 20kg, e.g. 190mm blocks) — Ensure unit weights and sizes of materials are reduced to acceptable levels where manual handling is unavoidable — Specify easily achievable tolerances where possible — Specify lifting hooks (e.g. coping stones, large masonry items)		
Cuts and abrasions	— Avoid specifying materials and components with sharp edges, corners, etc. where reasonably practicable	*This is more of a trade capability and training issue, e.g. using brick ties*	
Carcinogenic diseases	— Avoid specification of known carcinogenic materials and substances — Where no alternative, ensure that adequate information is available at tender stage		
Injury to eyes	— Operations involving splashing		
Respiratory injuries	— Avoid specifying materials and substances which are likely to cause respiratory problems where possible, e.g. epoxies — Design to avoid cutting, chasing, etc.		
Deleterious materials	— Check if allowed in your appointment terms — Check if acceptable to client's building insurers		

Hazard Awareness and Risk Identification Checklist (HARI)

Risk	Designers to identify and consider the following significant risks and other factors – SFARP	Comments, Actions, References	RAG Colour
Skin diseases	— Avoid specifying materials and substances which are likely to cause skin diseases (e.g. dermatitis) where possible; protective clothing necessary for cement and lime mortars.	*Protection against exposure to wet cement and mortar is a trade issue*	
K	**Cladding/glazing** Flat roofwork, pitched roofwork, masonry, brickwork, blockwork, stonework, panels, windows, patent glazing, sheeting, tiling, slating		
Temporary instability	— Avoid designs which involve temporary instability during construction, or specify an erection sequence that avoids it. If unavoidable, detail temporary support measures required — Temporary fixing of windows/curtain walling balustrades and guard rails	*This may be a specialist subcontractor designer issue*	
Falls from height	— Maximise prefabrication, adopt simple details and allow for early installation of floors, roof decks, stairs, parapets, permanent edge protection etc. to minimise risk from high-level working — Specify easily achievable tolerances where possible — Detail to allow easy connection of safety lines, harnesses etc. where necessary — Use large decking, cladding panels, domed roof lights — Consider future maintenance and cleaning, especially balconies — Consider window cleaning from inside where possible — Consider permanent access or fastenings — Consider appropriate type of temporary and permanent edge protection to roofs — Window restrictors, handle accessibility, cill heights and guarding	*BS 8560:2012 Code of practice for the design of buildings incorporating safe work at height*[30]	

Hazard Awareness and Risk Identification Checklist (HARI)

Risk	Designers to identify and consider the following significant risks and other factors – SFARP	Comments, Actions, References	RAG Colour
	— Consider heights of balustrading where publicly accessible, or where seating is provided adjacent (e.g. food courts)		
Construction loadings	— Identify construction loadings on drawings for mechanical installation plant and temporary works allowances and stacking of materials		
Falls through fragile materials	— Avoid specifying fragile materials (e.g. roof-light panels) — Consider installation, fragility and glazing of roof lights — Provide guard rails around roof lights or raise up		
Falling objects	— Ensure adequate lifting provisions on components — Maximise prefabrication — Safe access for future maintenance and cleaning of facades — Review specification for temporary fixing of windows/curtain walls to avoid being blown out by gusts of wind before being permanently fixed (cause of two PI notifications) — Design out complex fixing details of large elements at high level with small components — Ensure no gaps in balustrading where objects can pass through above public areas, e.g. atria, transport hubs, etc. — Advise contractor of need to tether tools, elements and materials, where working above others	BS 8560	
Fire	— Specify non- or low-flammable and non-combustible materials and products where possible — Consider the end use of the building, e.g. kitchen, and check with the client's building insurers for extra requirements — Consider escape from roofs in a fire		

Hazard Awareness and Risk Identification Checklist (HARI)

Risk	Designers to identify and consider the following significant risks and other factors – SFARP	Comments, Actions, References	RAG Colour
L	**Furniture, finishes and equipment** Stone, ceramics, coatings, paints, sealants, adhesives, wood, wood-based materials, synthetic materials		
Hazardous materials/ substances	— Avoid specifying finishes involving hazardous materials/ substances where reasonably practicable — Substitute safer alternatives — Specify pre-finished components where reasonably practicable		
Dust/fumes	— Avoid specifying surface preparation, application methods and processes likely to release hazardous dust or fumes where reasonably practicable, i.e. cutting, drilling, abrading, polishing, etc. (e.g. with solvent paints, adhesives and spray paints) — Avoid dust from creating site-mixed powder materials		
Noise/vibration	— Avoid specifying finishes requiring use of vibrator tools or noisy equipment for surface preparation or application methods where reasonably practicable		
Fire	— Avoid specifying materials using inflammable solvents		
Musculo-skeletal injuries	— Size of furniture and components such as glass screens, reception desks, sliding folding partitions to be considered in terms of vertical access via goods lifts, hoists or at last resort stairs if of man-handleable size. Horizontal access also to be considered e.g. trolleys, skates, etc.		

Hazard Awareness and Risk Identification Checklist (HARI)

Risk	Designers to identify and consider the following significant risks and other factors – SFARP	Comments, Actions, References	RAG Colour
M	**Risks to building users and visitors**		
Crushing	— Escape routes from loading bays — Protection of pedestrians from vehicular routes — Boundary walls designed to fall away from railway tracks, roads, playgrounds and pedestrian areas — Vehicle barriers to prevent runaway cars falling from height in icy conditions or where driver loses control		
Drowning	— Hazard warning signs to be provided, e.g. 'Danger Sudden Drop'. Life buoys? — Railings to be 1.1m high and difficult to climb. Designed to resist impact by vehicles — Whoever takes responsibility for the maintenance of river walks is to take the decision as to the appropriate level of protection to meet the duty of care requirement of health and safety legislation. They must also put into place a management plan for the maintenance of such protection		
Fire	— Escape for deaf or hearing-impaired people, particularly in public buildings – consider vibrating alerts or visual alarms — Consider evacuation as well as access and advise employers/tenants of their duty to provide a suitable evacuation strategy for a wide range of users and to carry out fire risk assessments	*These should be included in the Fire and Inclusive Design Strategies under parts B & M of the Building Regulations*	

Hazard Awareness and Risk Identification Checklist (HARI)

Risk	Designers to identify and consider the following significant risks and other factors – SFARP	Comments, Actions, References	RAG Colour
Collision/ trapping	— Automatic doors – risk assessment to BS 7036 — Automatic gates – Gate Safe to be referenced	BS 7036 Code of practice for safety at powered doors for pedestrian use[31] Gate Safe[32]	
Falls from height	— Operation of high-level blinds — Window controls as Part N, less than 1700/1900mm above floor level — Fixing to be provided for ladders more than 6m long — Horizontal force on balustrades specified as appropriate for use, e.g. public use – refer to AD K — 100mm gaps between balustrades if children under 5 are likely to be present — Glass balustrades and full-height glazing at height – consider risk if toughened glass shatters — Safety glazing to Part N and BS 6262 — Restrictors and stays to opening windows to prevent accidental falls and children climbing out — Ability to attach safety harnesses to gantries and walkways, and anchor lines — Guarding to level changes, riverbanks, ditches, stairs, under-stair soffits, ramps	BS 8560 Code of practice for the design of buildings incorporating safe work at height[30] Building Regulations, Approved Document N & AIF Information Building Regulations, Approved Document K[33]	
Musculo-skeletal injuries	— Outward-opening doors and windows – walking into leading edges, especially at head height		

Hazard Awareness and Risk Identification Checklist (HARI)

Risk	Designers to identify and consider the following significant risks and other factors – SFARP	Comments, Actions, References	RAG Colour
Burns/scalding	— Low surface temperature radiators where toddlers or elderly may be present. Also avoid unprotected pipework — Flue outlets located away from public areas or protected		
Drowning	— Full risk assessment, particularly for unsupervised pools — River walkways etc.		
Safe by Design/ rape/attack	— Consider personal safety and sense of safety — Advice of crime prevention officer or guidance to be sought — Good surveillance, lighting, minimisation of recesses	*Safe by Design, Local Crime Prevention Officers, BREEAM[26] advantages*	
Slips/falls/ breaks and bruising	— Slip resistance of floors specified to entrance, corridors, swimming pools and sanitary accommodation, especially in wet conditions. Entrance matwells and slip resistance research related to cleaning materials used — Good extraction ventilation to kitchens, bathrooms and swimming pools — Undulations and paving trip hazards — Guarding against falls in level, steep slopes — Stair and step nosings in contrasting colour with good lighting, including to external stairs and fire escapes — Protection to features that may attract children, e.g. skateboard ramps, features to climb	*HSE Slips assessment tool www.hse.gov.uk/ SLIPS/sat/index.htm*	

Hazard Awareness and Risk Identification Checklist (HARI)

Risk	Designers to identify and consider the following significant risks and other factors – SFARP	Comments, Actions, References	RAG Colour
N	**Maintenance/repairs**		
Falls from height	— Consider ease of replacement of light bulbs, especially above stairs or in high spaces, atria, etc. — Design in adequate safe systems of access, edge protection, provision for the attachment of safety equipment, eyebolts, rails etc. where necessary — Base for ladders provided as in Approved Document N — Position controls, valves and equipment requiring regular maintenance at low level (or lowerable) — Low-maintenance equipment/fittings where practicable — Design process followed to select suitable mansafe/latchways system if no alternative. Harnesses and training for use included in specification — Cat ladders and walkways designed to Building Regulations and relevant BS — Handrails to be specified near lift-up access hatches	*Provide information regarding built-in safety facilities, etc. in H&S file/ maintenance manual*	
Falls	— Steep slopes, including landscape maintenance		
Falls through fragile materials	— Avoid specifying fragile materials where possible, but may exist — Design in adequate safe systems of access, edge protection, provisions for the attachment of safety equipment, etc. where necessary	*Provide information regarding fragile materials and inbuilt facilities in H&S file/ maintenance manual*	
Electrocution/ scalds	— Provide adequate isolation facilities for all plant and equipment		

Hazard Awareness and Risk Identification Checklist (HARI)

Risk	Designers to identify and consider the following significant risks and other factors – SFARP	Comments, Actions, References	RAG Colour
Manual handling/ Musculo-skeletal injuries	– Design components for ease of handling and replacement – Provide adequate access facilities, working space and lifting facilities around all plant and equipment where necessary – Large roof hatches with hydraulic or mechanical assistance to open and close	*Provide information regarding access and lifting facilities in H&S file/maintenance manual*	
Hazardous materials/ substances	– Avoid hazardous materials and substances – Provide information on existing or unavoidable hazardous materials or substances, e.g. asbestos, lead paint		
P	**Dismantling/demolition/future alteration/refurbishment**		
Uncontrolled collapse	– Provide information regarding design parameters, design loadings, means of ensuring structural stability, construction details, specific alteration/demolition hazards (i.e. pre-stressing, suspension, cantilevers etc.)	*Designers do not need to tell demolition professionals how to demolish all buildings, only identify significant risks*	

Table 6.2 Zero-HARM hazard awareness and risk management register

Once the relevant risks for the project have been identified they should be added to the drawings, and if considered proportionate and necessary for convenience of the team, collated into a risk register for ease of reference. BIM and other electronic drawing techniques should do this automatically without need for repetition, reducing the risk of error. This also helps when issues are revised.

Zero-HARM Hazard Awareness and Risk Management Register (0-HARM)						
Project Title		Project No.		Risk Tolerability	Risk not tolerable	
Prepared By		Checked By			Further consideration req'd	
Date		Stage			Risk tolerable	
Risk Ref.	Relevant Hazard/ Risk Identification	Design Control Measures/Residual Risk/Drg. Ref.		Action Owner	Date Required	Completed

CDM 2015 and the use of red, amber and green lists

The following red, amber and green lists are available on the HSE website and are part of the CDM 2015 Industry Guidance.[34] They are practical aids for designers on what to eliminate and avoid as well as what to encourage. For instance, examples such as 'glazing that cannot be accessed safely' falls under the red category, which should be 'eliminated from the project where possible' whereas 'off-site fabrication' falls under the green category, as it will positively improve safety on site and should be encouraged. This list can be used to inform the identification of the level of risk in the tools above.

While they form a useful reference point during design review, they should be considered in the collaborative team discussions along with all other factors SFARP. A commentary is provided of the kind of discussions these lists generate.

Table 6.3 Red, amber and green lists[34]

HSE red, amber and green lists

Red Lists: Hazardous procedures, products and processes that should be eliminated from the project where possible	Author's comments
— Lack of adequate pre-construction information, e.g. asbestos surveys, geology, obstructions, services, ground contamination etc. — Hand scabbling of concrete ('stop ends', etc.) — Demolition by hand-held breakers of the top sections of concrete piles (pile cropping techniques are available) — The specification of fragile roof lights and roofing assemblies — Processes giving rise to large quantities of dust (dry cutting, blasting etc.) — On-site spraying of harmful substances — The specification of structural steelwork which is not purposely designed to accommodate safety nets — Designing roof-mounted services requiring access (for maintenance, etc), without provision for safe access (e.g. barriers) — Glazing that cannot be accessed safely; all glazing should be anticipated as requiring cleaning and replacement, so a safe system of access is essential — Entrances, floors, ramps, stairs and escalators etc. not specifically designed to avoid slips and trips during use and maintenance, including effect of rain water and spillages — Design of environments involving adverse lighting, noise, vibration, temperature, wetness, humidity and draughts or chemical and/or biological conditions during use and maintenance operations — Designs of structures that do not allow for fire containment during construction	The statement 'should be **eliminated** from the project where possible' needs clarification, i.e. where this is not possible due to other influencing factors these issues need to have suitable risk control measures imposed **SFARP**

HSE red, amber and green lists

Amber Lists: Products, processes and procedures to be eliminated or reduced as far as possible and only specified/allowed if unavoidable. Including amber items would always lead to the provision of information to the principal contractor	Author's comments
— Internal manholes/inspection chambers in circulation areas — External manholes in heavily used vehicle access zones — The specification of 'lip' details (i.e. trip hazards) at the tops of pre-cast concrete staircases — The specification of shallow steps (i.e. risers) in external paved areas — The specification of heavy building blocks i.e. those weighing >20kgs — Large and heavy glass panels — The chasing out of concrete/brick/blockwork walls or floors for the installation of services — The specification of heavy lintels (the use of slim metal or hollow concrete lintels being alternatives) — The specification of solvent-based paints and thinners, or isocyanates, particularly for use in confined areas — Specification of curtain wall or panel systems without provision for the tying (or raking) of scaffolds — Specification of blockwork walls >3.5 metres high using retarded mortar mixes — Site traffic routes that do not allow for 'one way' systems and/or vehicular traffic segregated from site personnel — Site layout that does not allow for adequate room for delivery and/or storage of materials, including specific components — Heavy construction components which cannot be handled using mechanical lifting devices (because of access restrictions/floor loadings etc.)	The statement 'eliminated or reduced as far as possible and only specified/allowed if unavoidable' – again further explanation is required. Where 'unavoidable' due to other influencing factors these issues need to have suitable risk control measures imposed SFARP

HSE red, amber and green lists

— On-site welding, in particular for new structures — Need to use large piling rigs and cranes near 'live' railways and overhead electric power lines or where proximity to obstructions prevents guarding of rigs	
Green Lists: Products, processes and procedures to be positively encouraged	**Author's comments**
— Adequate access for construction vehicles to minimise reversing requirements (one-way systems and turning radii) — Provision of adequate access and headroom for maintenance in plant rooms, and adequate provision for replacing heavy components — Thoughtful location of of mechanical/electrical equipment, light fittings, security devices etc. to facilitate access and away from crowded areas — The specification of concrete products with pre-cast fixings to avoid drilling — Specify half board sizes for plasterboard sheets to make handling easier — Early installation of permanent means of access, and prefabricated staircases with handrails — The provision of edge protection at permanent works where there is a forseeable risk of falls after handover — Practical and safe methods of window cleaning (e.g. from the inside) — Appointment of a temporary work coordinator (BS 5975) — Off-site timber treatment if PPA- and CCA-based preservatives are used (Boron or copper salts can be used for cut ends on site) — Off-site fabrication and prefabricated elements to minimise on-site hazards — Encourage the use of engineering controls to minimise the use of personal protective equipment[35]	These are recommended good practices but sometimes circumstances will mitigate against their provision or are beyond the control of designers

Risk and hazard logs that are fit for purpose

On the following pages are some examples of simple and more complex approaches to logging risks, as well as a more visual approach.

Design Hazard Management Register							
Project & No:	A Big Building		12345	**Work Stage:**		**Revision & Date:**	Rev. A	4 April 2013
Risk ref. no. and date opened	Element or activity and hazard or potential to cause harm	Persons at risk, their operations and the consequences	Design team risk reduction proposals required to mitigate the risk (with options/alternatives recorded)		Responsibilities & Actions		
				Required date of action or work stage	Risk closed, open or residual	Risk action owner
Unique no. to be added and date	Significant hazard to be entered This could be a design element (01), material (02) or activity (03) Trivial issues not to be included	Identify all persons exposed to the hazard, the work undertaken and the effects on health	Design team actions, with options considered, required to eliminate or reduce the risks. If none possible, what information will be passed to the contractor to control the risk?	Project gateway to be entered into	Enter current status of design development	Initials of relevant discipline as inserted into footer
01 06/04/07	Existing underground services Major gas, sewer, water and electrical supplies known to cross the site	Injuries to construction workers during groundworks and excavations, e.g. electrocution, explosions, falls	Client to commission adequate survey of existing services as early as possible so the design team can coordinate site and building layout to minimise the diversions or excavations. Detailed survey drawings to be made available to contractor at tender stage	Prior to Stage C	Open	Client, services engineer, architect
02 06/04/07	Asbestos Materials likely to be present in existing building	Serious ill-health to all persons working on, visiting or passing site during project from feasibility to completion	Asbestos survey of appropriate type for project to be commissioned by client. Design team to review and integrate an asbestos management regime with the design, phasing and construction proposals. Residual asbestos statement required for inclusion in health and safety file	Prior to tender	Closed	Client, architect, asbestos specialist, construction expert
03 06/04/07	Cleaning and maintenance of fragile glass roof to atrium	Maintenance workers and users during cleaning operations Falls from height or being struck by falling objects	Design team to integrate cleaning and maintenance regime proposals with development of envelope and form of atrium. All details of system and operation to be included in health and safety file	Prior to Stage D planning application	Residual	Architect, structural engineer, cladding specialist
Client	Client		Architect	Struct. Eng		Services Eng	P. Contractor
Consultants	PM		CDM–C	Landscape		Cost Consultant	

Table 6.4 A simple risk register

The simplicity of a risk register (as above) is important; trying to convey too much information in a narrative way does not encourage a collaborative and interactive approach, especially in project team meetings. The example above may be suitable as a record of agreements and for confirmations of approach.

PART A – HAZARD ELIMINATION/REDUCTION

A	B	C	D	E	F			G	H		
Ref.	Specific Location/ Activity	Phase	Author – Name & Company	Potential Hazards	Initial Risk Rating			Action by Designer to Eliminate/Reduce Risk Rating	Residual Risk Rating		
					L	S	R		L	S	R
STRUCTURES											
S.1	S01 Piers	C,M,D	TAN	Working in proximity to live traffic	4	5	20	S01 crosses several major roads and impossible to configure within alignment constraints to avoid constructing near live traffic. Risk cannot be eliminated or significantly reduced. Position piers as far away from live traffic as possible. Single span	3	5	15
S.2	S01 Working in A2 c.r. adjacent to live traffic	C,M,D	TAN	Working in proximity to live traffic	4	5	20	S01 crosses several major roads and impossible to configure within alignment constraints to avoid constructing near live traffic. Cannot be eliminated and therefore no mitigation available to designer	4	5	20
S.3	Pier bearing installation, maintenance & replacement (S01, S04, S10(N) & S10(S)	M	TAN	Working in proximity to live traffic Working at height	5	5	25	Design as integral bridges without bearings considered but structure too long for this. No alternative mitigation available to designer in this respect	5	5	25

Table 6.5 A complex design hazard log
This is an example of over-complication in a design hazard log. This may be suitable for construction phases of complex nuclear power stations, railways, highways or petro-chemical installations, for example. However, this is not a proportionate response for most projects, even large ones, at the design stage. If required this may be more suitable for the principal contractor to produce to manage normal contractor risks on site. The CDM Regulations do not require this type of information, especially at design stages where it can effectively detract from safe design.

	PART B – TRANSFER OF INFORMATION			PART C – CONTRACTOR IMPLEMENTATION			
	J	K	L	M	N	O	P
Information Provided About the Residual Hazards – Drawing/Document	Design Manager Responsible – Name	Designer/Constructor Discussion Date & Comments	Status Active/ Closed	Construction Manager Responsible – Name	Control Measures Required	External Review of Control Measures? Y/N (by whom)	Control Measures Identified in:
STRUCTURES							
Note on drawing – method statement required to cover particular issues relating to the complexity of the existing slip roads and underneath and requirements for temporary road closures for pier construction	TN	17/05/06 & 21/05/06	Active	MB	Works to be carried out with lane closures in accordance with TM phases. Provide protected safety zone with barriers		
Note on drawing – method statement required to address the particular difficulty of working on an island site	TN	17/05/06 & 21/05/06	Active	MB	Works to be carried out with lane closures in accordance with TM phases. Provide protected safety zone with barriers		
Note on drawing – method statement required addressing handling and installation of heavy components at height with restrictive clearances. Log in H&S file	TN	17/05/06 & 25/05/06 Residual risks to be noted in H&S plan – barriers & TM required	Active	MB	Provide protected safety zone with TVCBs to mitigate traffic risk		

Safety design reviews

Ideally the designer should try to identify the key CDM issues and challenges on their drawings. This is to enable identification and discussion in a collaborative way at future design team or CDM review meetings, on larger projects. The HARI checklist and 0-HARM register above could both be used to organise the thinking process and provide suitable and sufficient information and proposed solutions.

When the safety issues of a design proposal are reviewed in meetings it is very important to facilitate a thorough and comprehensive discussion. The provision of visually clear and intelligible drawings, models, 3D images, sketches and so on is vital for clearly and immediately identifying and recording discussion points and ideas. For example, flipcharts are a simple but effective tool to help communicate an issue and its proposed resolution to all in the room. They can be photographs and prints given to participants for inclusion in their design development processes well in advance of the production of formal minutes, which often do not capture the full extent of the discussion. Emerging technology such as handheld devices will offer many other solutions. By circulating terms of engagement to all meeting attendees in advance, open discussion will be encouraged between project team members.

Terms of engagement for safety design reviews
Environment:
— Who needs to attend the meeting or review? Only have people in the room who will contribute, add value and know their role in the meeting. This should be the whole project team, not just designers
— Timescale to be agreed in advance to allow sufficient time to review the risk and hazards thoroughly
— Ensure the meeting or review has everyone's full attention. Mobile phones are to be switched off
— Location of the meeting or review. Everyone should attend in person or via video conference. Does VC provide the maximum value?
— Capture of information. How are the issues, risks and hazards captured for visual communication to others?

- No blame culture. All views are valid
- Room should have sufficient space for displaying any material that needs to be discussed
- Drawing aids, smart board, monitors for BIM models

Meeting room format:
- Desks or no desk, chairs or no chairs? No desks opens the room to conversation, no chairs keeps the meeting short
- Use of flipcharts, photos, interactive whiteboards? Encourage drawing and capture and display the outcomes during the meeting. Circulate outcomes immediately after
- Would the use of technology such as iAnnotate PDF work in this environment for rapid distribution of ideas, or Dropbox?

Timing:
- Hold reviews and meetings at key design hold points. Select when buildability reviews are most appropriate. (Suggested around RIBA Stage 2/ Stage 3 before elements are fixed by statute)
- Regular reviews can be integrated seamlessly with normal design team meetings

Encourage open discussions:
- Reasonably practicable (SFARP) solutions by collaboration; minimising risk, maximising safety & design quality
- Agree tolerable buildability, maintainability and usability
- Risks to be 'sufficiently and suitably' managed
- Mutual trust in the team
- Working to shared project goals – not individual/ company targets
- Speaking openly without fear of retribution, ridicule or reprisal
- Looking at the project life cycle – brief, build, operate, maintain and decommission
- Subcontractor input – would you design it differently if you knew this earlier?
- Trust – profit is expected to be made by all parties
- Listening as well as talking
- Big-picture thinking

Discourage:
- Hypothetical risk discussion
- 'Don't do it like that, do it like this!' culture
- Basing decisions around cost of design input only
- Pride and ego; reluctance to change

Watch out for barriers to open discussion:
— Lack of understanding of the 'big picture' or design development work that has gone on before the meeting (how did the designers get to where they have?)
— Intellectual properties/commercial advantage
— Design fees spent and signed off
— Lack of trust within the meeting

Minimising rather than eliminating risk

Working at height and cutting masonry are seen as some of the most dangerous practices in construction, and there is often a call to eliminate them completely so as to avoid high-risk activity. However, these are often essential on building projects, and it is not feasible to take them out of the equation. There have been recent attempts by the industry to find a different approach to this dilemma using creative solutions, as explained below.

In her 2012 speech to the Institute of Mechanical Engineers (IMechE), Judith Hackitt, the Chair of the HSE, spoke about 'Engineering safe sustainable solutions – from cradle to grave' which encapsulates the issue.

> I believe we need to be clear that risk management and innovation are entirely compatible. Risk elimination and innovation, however, is an oxymoron.

We are all familiar with the precautionary principle. In its most generic sense we all utilise it all of the time in our daily lives – weighing up the risks of what we do or are about to do versus the benefits. But in its formal definition and subsequent application there seems to be little doubt that the precautionary principle can be a barrier to innovation.

But the basic facts are these:

There will nearly always be a threat or risk of harm of some magnitude; and there will almost never be full unequivocal scientific knowledge about the situation. Building public confidence will not come from telling people that 'we know best'. What will help to deliver it is:

— *acknowledging justifiable fear or apprehension of the new and unknown;*

> — *explaining innovations in terms of benefits and risks;*
>
> — *being honest about what can be done to reduce but not eliminate risk; and*
>
> — *constantly reminding people that no action is by any means risk free.*
>
> Risk communication is also about differentiating between real risk and risk-averse behaviour.[36]

We must therefore all realise that in order to design creatively, allowing for innovation but incorporating acceptable levels of safety is a delicate balance. We must combine health and safety knowledge and its coordination in the context of the other constraints and influences of the project.

Can we realistically eliminate working at height?
A typical example of this issue is included in BS 8560 Code of practice for the design of buildings incorporating safe work at height, written by a team of construction industry professionals, architectural designers (including the author), industry experts and HSE working-at-height specialists. From very different backgrounds this team developed a mutually collaborative approach to assist designers with incorporating safe working at height measures into their designs. This was perhaps the first important document to accept that elimination of working at height is frequently not possible but that providing safe methods, or installations and equipment, to suit the individual circumstances of each project is essential. The code includes the following introduction:

> *This British Standard encourages designers to assess, as early as possible in the design process, how work at height can be minimized, and where required the provision of practical, efficient, cost-effective solutions for the safety of those who work at height. For the purposes of this British Standard, the term 'designer' covers either an individual engaged in design or a team, perhaps representing several disciplines. An integrated design team offers benefits in terms of collective design consideration, including knowledge, experience and problem solving. New-build and refurbishment projects require people to work at height over the lifespan of a building in order to construct, clean, maintain and repair it.*
>
> *By its nature, work at height is hazardous and presents the risk of a fall. Falls from height account for a significant proportion of fatalities and serious injuries experienced during construction and maintenance. As a significant safety risk, it is important that everyone, including designers, who*

work within the construction industry, give it the appropriate attention. To reduce the number of accidents, designers have duties under health and safety legislation, so far as is reasonably practicable, to avoid the need for work to be carried out at height. Part of a designer's duty is to adapt the design where this cannot be achieved so that equipment can be provided to prevent falls. If provision of fall prevention equipment is impractical, equipment or systems of work to minimize the distance and consequences of a fall can then be included within the design.

By working collaboratively, the design team is in a strong position to make provisions within their designs that reduce the risk of falls occurring. Designers assess and manage many competing factors as they prepare their designs. At the concept design stage the form of the building develops and this can be influenced by factors such as function, location, aesthetics, cost and planning. In addition, there are other factors such as building and fire regulations, sustainability, buildability and maintainability to be addressed. The best time for designers to consider work at height is during the early stages of the design: how it can be minimized and carried out in a manner that provides an appropriate level of safety. This is the focus for this British Standard. By early consideration of the extent, nature, duration and frequency of work to be done at height, appropriate equipment and techniques for use in construction, cleaning, maintenance and repair can be identified. Also how access to and from equipment and places of work at height, and the loads equipment imposes on the structure or surrounding ground can be included early in the design process.[30]

DIOHAS have developed case studies which try to support this more creative approach to minimising risk when working at height. With reference to current legislation and guidance from CIRIA they have produced quick-reference storyboards that guide you through current guidance, risks and potential solutions.

GUIDANCE – Flat Roof Maintenance Access Options

Flat Roof Maintenance Access

Permanent ramps, staircases, walkways and platforms with full edge protection are generally by far the safest. Although they may not always be seen as practicable, the reasons for this must be vigorously challenged, as carrying equipment, tools or even notebooks is very difficult. Lateral thinking should be employed to find a way to incorporate them.

Permanent, fixed ladders (whether vertical or inclined) should be used only rarely, when other options are impracticable or as an escape only alternative. Active consideration should be given to the provision of a permanent latch-way rail to facilitate locking on during use.

Ladders must be limited in length with rest platforms and a platform should be provided where there is a foreseeable need to carry out a task such as operating a valve or opening an overhead trap-door. Each particular scenario must be assessed on its own merits and the specific requirements of the Work at Height Regulations 2005 complied with.

CIRIA C686 – Access for Maintenance & Repair

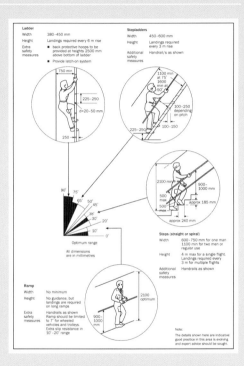

SCOTT BROWNRIGG / DIOHAS – PROPORTIONATE AND PRACTICABLE CDM FOR DESIGNERS

Figure 6.1 DIOHAS case study exploring flat roof maintenance options

CASE STUDY – Strategic design – residential façade access system

The problem/challenge
Landlord controlled window cleaning access to high value flats with balconies to main elevations. Impractical to use a suspended access cradle due to the need to climb out of cradles and over balconies. Access via flats not acceptable to tenants for security reasons. Access from common parts not possible, and balconies not continuous for security reasons.

The risks
Falls from height during window cleaning operations.

The solution
Roped access solution for operatives to access each balcony area from which safe cleaning operations can take place.

The benefits
Landlord can ensure all **windows cleaned at regular intervals**. Operatives clean most windows from safe balconies.

Key points
IRATA registered company consulted on rope attachment design and details. **IRATA trained rope access operatives employed** to ensure safe systems of working. Highly specific project details justified a roped access solution, which would need to be equally justified if proposed on other projects.

Full length but separated balconies

IRARA guidelines used

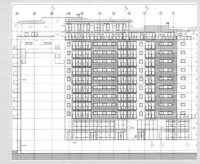
Section and elevations analysed for appropriate systems

SCOTT BROWNRIGG / DIOHAS – PROPORTIONATE AND PRACTICABLE CDM FOR DESIGNERS

Figure 6.2 DIOHAS case study exploring roof-light installation

Figures 6.3 and 6.4 Intentional cutting of paving slabs to allow for a slot drain, and building margin. Dimensional and on-site coordination required, but not necessarily resolved fully on the drawing. Contractor to be advised to make allowances for on-site cutting with appropriate dust prevention and personnel and public protection measures

Can we realistically eliminate cutting of blocks?
Cutting and laying paving and blocks is a generic trade-related process. However due to the often uncontrolled cutting of blocks, bricks, paviors and stones by operatives with high-powered circular saws, large amounts of silica dust is released into the atmosphere.

This not only affects the immediate operatives but their colleagues on site, other site visitors, neighbours and the local population.

It is unreasonable to assume that the use of blocks can be eliminated, but designers can help to mitigate these effects by minimising the amount of cutting required and setting out coordination proposals. However, it is disproportionate to eliminate cutting completely. For instance, coordination and communication between the designer and the contractor perhaps in terms of setting out could minimise the amount of cutting required.

By interacting with the Construction Dust Partnership (see Appendix I) a proportionate designer approach can be understood. This is a classic example of how collaborative working by all design and construction participants can contribute to good CDM outcomes. Figures 6.3 and 6.4 on the previous page are an example of good coordination. A slot drain is to be created so cutting of blocks diagonally is an essential design intent. This could be an on-site or off-site activity but the issue is raised at pricing stage.

Designers are tasked with identifying the minimum amount of cutting required (SFARP) to achieve the design intent from a CDM safety as well as a cost perspective, and must consider the methods to use to achieve this. Contractors are then required to identify their proposed methods to control this, in principle, at tender or contract stages. A visual methodology could explain the construction stage intent succinctly to all project team members from the designer, cost controller and site operative to the chief executive, without the need for complex narrative paperwork where risks may be lost.

Two further examples of tools for problem solving follow. Figure 6.5 is a case study created by DIOHAS, in association with the RIBA Regulations and Standards Group. These are mostly targeted at capturing good practice, but also demonstrating design lessons learned from accidents. Figure 6.6 is a matrix which demonstrates how we can visually communicate the project team's design and construction intentions.

CONTRACT STAGE DESIGNER INPUT – Plasterboard – musculo-skeletal injuries

The problem/challenge
Lack of information about board weights for operatives. Wide range – from jobbing builders to specialist drylining contractors – but **mainly subcontracted site operatives injured. Sites driven by cost** – not by considerations of health and safety – speed rules! Board handling is part of the project delivery, and to be considered by all stakeholders.

The risks
Operatives are taking excessive risks lifting boards. Operatives treat musculo-skeletal injury as a risk that you take and that can't be avoided. Musculo-skeletal damage is occurring without being identified. Operatives have a **shortened working life**.

The solution
Boards to have **heavy lifting symbols attached**. Progress through working together to find approaches that work. The **designer can encourage good working practices**, mechanical moving, lifting aids, hoists, goods lifts, drawing specifications, and red, amber, green lists.

The benefits
Better plasterboard installations with **more motivated operatives, and increased productivity. Longer working life** of operatives and skills retention. Reduces risk to companies against future injury claims.

Key points
Work with the stakeholders to jointly develop an approach to reduce the risk of musculo-skeletal injury. Safeguard the health of individuals working with boards by **mechanical aids or good lifting practices**.

Half boards and equipment

SCOTT BROWNRIGG / DIOHAS – PROPORTIONATE AND PRACTICABLE CDM FOR DESIGNERS

Figure 6.5 DIOHAS case study about lifting boards

< Work Stage: Insert name >				Project: < Insert >					Date: < Insert >	
		Project Preparation Matrix			**Hazard Awareness & Risk Management**				< Insert Trade package >	
Ref. no.	HAZARD Material, hazardous activity, location or circumstance	SIGNIFICANT RISKS Generic trade & project-specific risks Consequences		PERSONS AT RISK	ELIMINATE or AVOID risks SFARP Mitigation measures > (early stages)	REDUCE or MINIMISE risks ALARP (During all design stages) Prioritised safe systems of work options available to resolve			INFORMATION To be provided with the design	OTHER SPECIALIST GUIDANCE & COMMENTS Health issues Further references
1.0	**Block paviors**	Type of risk		Individual or multiple	Do not use?	Design method option – 1	Design method option – 2	Design method option – 3	Other project documentation	Further information
e.g.		Generic/trade-specific risks are those that should be mitigated by trained and experienced tradesmen		Many people can be affected by dust and noise, whether operatives, supervisors, visitors, neighbours, or the wider environment	This may not be a viable option ✗	Use of designer skills to achieve visual intent with minimisation of harmful potential	Use of designer skills to achieve visual intent with minimisation of harmful potential	Use of designer skills to achieve visual intent with minimisation of harmful potential	Refer to: Project drawings cross referenced Specification Cost plan	*RR878* Levels of respirable dust and respirable crystalline silica at construction sites[37] Signage *Specialist design or client input, climatic conditions Costs/benefits*
1.1		Musculo–skeletal injuries during laying		Operatives	✗	Reducing size of units or elements	Encourage mechanisation by using large units *Climatic conditions or space issues may prevent this*		Access, size of project and duration for mechanisation will dictate the methods chosen. Small refurbishments may not justify the use of large plant	*RR878*
1.2		Respiratory risks during cutting		Operatives Other workers Site staff Neighbours Public	✗	Minimise cut blocks in paving pattern	Ensure		Containment, suppression, etc.	On-site availability of suitable cutting equipment and containment to cutting areas is essential	*RR878*
Client		**Client**		Architect		Struct. Eng		Services Eng		P. Contractor
Consultants		**PM**		CDM–C		Landscape		Cost Consultant		

Figure 6.6 A visual hazard awareness and risk management matrix for cutting blocks, demonstrating a design risk management approach with SFARP

7 Conclusions: CDM Differently

While compiling and collating articles, learned papers and research documents for this guide it has become increasingly clear that over the last 20 years the intentions of CDM have been misinterpreted by many, inside and outside the industry. It is clear that the motivations of the various duty holders and legislators trying to implement this regulatory framework are generally altruistic, moral and ethically founded. However after two decades the Health and Safety Executive has realised that the tone and format of the Regulations has tended to set up an unnecessarily complex legislative culture, which has spawned ancillary compliance and competence activities creating an illusion of safety rather than the actual improvements in safety they hoped for.

Essentially, as every concept design is established it is important to identify and highlight the key CDM challenges on the project. Some of these are architectural features or associated activities (rather than risks or hazards) that are unusual, difficult to manage, transient, not visually apparent on drawings or lost in other information. However, these constitute key elements of the design intent as much as the client brief, not to be eliminated lightly. Once the whole industry accepts this approach we can hope to have a collaborative environment, within which CDM and health and safety can have a proportionate and practicable impact on architectural design.

While trying to reduce accidents on site and draw in latterly disengaged client and designer communities, a 'disproportionately applied monster' has been created. This was recently reported to the CIC Health and Safety Expert Panel (November 2014) by Philip White, HSE Chief Inspector of Construction.[38] He also expressed a view that 'the design professions will soon get up to speed on the role of "principal designer" when it is realised that current practices have gone overboard with paperwork; a much lighter touch and less-zealous approach is required and the whole process will be brought into perspective.' Architectural designers must be encouraged to exercise their creative skills and to push the boundaries of innovative design. However, in doing so, significant considerations of constructional safety and future maintenance should also be borne in mind, but proportionately.

Hopefully the CDM Regulations 2015 will lay sturdy foundations for this balancing act.

Appendix I
Professional competence, support and information

Professional competence

Professional competence is of the utmost importance in health and safety. However the architectural design community is besieged with pre-qualification questionnaires assessing competence from clients and organisations that do not have a comprehensive grasp of the issues outlined in this guide. This imposes an unacceptable burden on the designers and creates an illusion of safety that the HSE and professions are determined to eradicate. There clearly does need to be one professionally approved questionnaire that assesses a proportionate level of capability. Currently the Publicly Available Specification 91 (PAS 91), outlined in an extract below, is the leading and hopefully sole contender.

PAS 91: 2013 Construction pre-qualification questionnaires (PQQ)

PAS 91: 2013 replaced PAS 91: 2011. It provides a set of questions that can be asked by construction clients and buyers of potential contractors and suppliers as part of the pre-qualification process for construction projects. PAS 91 is aligned to the Government Construction Strategy, contributing to the target of achieving savings in construction costs of 15-20% through the reduction of tendering costs, and is set to be mandatory for Government projects from 2016. Seen as a significant improvement on the 2011 version, it includes questions to examine competence in Building Information Modelling (BIM) and collaborative information exchange in preparation for the Government 2016 BIM mandate (requiring fully-collaborative 3D BIM on public projects with project and asset information, documentation and data being electronic, as a minimum).

> *Once suppliers have completed one PAS 91 questionnaire, the same set of standard responses can be used by all buyers who are PAS 91 compliant. The format also allows suppliers to share their PQQ responses with other clients and buyers through PAS 91 online portals. PAS 91 could therefore be seen as a supplier's real-time portfolio of capability and as such, it requires constant updating to ensure that responses reflect current information.*[39]

Membership of Safety Schemes in Procurement (SSIP) or an affiliated organisation will help designers with the burden and provide advice on industry accepted levels of health and safety. The SSIP Forum was launched in 2009 as an umbrella organisation to facilitate mutual recognition between health and safety pre-qualification schemes.

There are a number of other organisations that assist designers with the implementation of CDM and health and safety, such as CONIAC, the CIC or DIOHAS, which are all listed later in this Appendix.

Card schemes for competency

There is much confusion about the need for and quality of the various card membership schemes.

From a professional designer perspective these cards solely demonstrate awareness of personal site safety while visiting active construction sites and are not a demonstration of construction safety knowledge, or knowledge of CDM issues. The CSCS card system is the most common, and is the subject of review.

> **Construction Skills Certification Scheme (CSCS) Cards**
>
> *Set up in the 90s, the scheme keeps a database of those working in construction that achieve (or can demonstrate they have already attained) an agreed level of occupational competence. Successful applicants are issued with a card giving them a means of identification and proof of their achievements. Today there are over 1.7 million cards in circulation, and they're increasingly demanded as evidence of occupational competence by many, including contractors, public and private clients. CITB administers the scheme under contract.*[40]

Industry support
HSE
The HSE often publishes lectures and papers on its

website that keep us aware of the latest government thinking on CDM.
www.hse.gov.uk

Construction Industry Advisory Committee (CONIAC)

> CONIAC advises the Health and Safety Executive (HSE) on the protection of people at work (and others) from hazards to health and safety within the building, civil engineering and engineering construction industry.
>
> Through this Industry based committee and sub-groups the HSE develop and report upon key H&S Construction Industry related issues for dissemination to the various constituent bodies and others in the industry. The CONIAC Committee have been appraised of and have participated in the development of CDM 2015 from the outset. CONIAC has four sub groups helping to take forward its work: Safety Working Group, Working Well Together Steering Group, Major Incidents, Health Risks Working Group.[41]

Construction Industry Council Health & Safety Panel (CIC)

The CIC provides a forum for health and safety representatives who are CIC members to meet, shape policy, discuss initiatives and develop responses on behalf of CIC. Membership of the panel includes the major chartered institutions, including the APS, ACE, ICE, IStructE, RIBA and HSE.

Key objectives:

> — To promote the consideration of health and safety in every sector of the built environment at all stages of projects: planning, design, construction, management, maintenance and demolition.
> — To identify and publicise examples of good practice.
> — To link and co-ordinate the activities of CIC members in relation to health and safety.
> — To represent CIC and its membership to key agencies such as the Health & Safety Executive, CONIAC and CSCS.
> — To promote health and safety risk management in all disciplines of education within the built environment.
> — To hold under review legislative developments in the UK and in Europe.[42]

Any relevant H&S and CDM issues can be brought to this committee via a participating institution's membership, for discussion and action.

Designers' Initiative on Health and Safety (DIOHAS)
www.diohas.org.uk

Aims:

> — Encourage clarity, common standards and understanding of health and safety issues amongst construction designers.
> — To focus, consolidate and disseminate proportionate health and safety information and procedures for the benefit of the industry and encourage the growth of a safer industry.
> — To achieve this through education, understanding, clear communications and enlightenment.
> — To establish and maintain a national standing as a recognised collective for the collation, wider understanding and dissemination of architectural and building design related health and safety issues.[43]

Relevant health and safety issues are presented, discussed and disseminated to members and other interested parties as part of the WREN Professional Indemnity Insurance Group.

Construction Dust Partnership Team

This organisation aims to protect construction workers against dust-related lung diseases by raising awareness and promoting the appropriate controls to prevent them, particularly for those undertaking high-risk tasks, and encouraging proportionate controls to minimise risk.

Further information

In order to understand the need for and methods of dealing with minimally bureaucratic team-based CDM approaches, we recommend you look at complementary professional sources, such as that for the ICE, below.

> 'Teamwork not Paperwork' encourages the reader to think differently, and then take action on the next project, which is the only way we can deliver better projects.[44]

Design Best Practice
www.dbp.org.uk

Appendix II
Healthy Design and Creative Safety – methods of teaching and learning about CDM

How can we embed good CDM practice into the education of design students?

In order to make CDM more appropriate and understandable for architects, it is essential that the process becomes embedded in design education as a creative and stimulating activity rather than a rule-based bureaucratic exercise. To try to explain this to industry and the legislators a research report was carried out by Sheffield University and a further symposium was held to explore and explain these issues.

In order to understand the context of CDM 2015 for aesthetic designer duty holders it is essential that these concepts are understood widely across the industry. The author chaired the CIC Education Task Group which facilitated the symposium and provided the following summary as a basis for change.

Summary
In 2012 the Construction Industry Council (CIC) Health and Safety Committee appointed a working group to report on the integration of health and safety into academic teaching in schools of architecture. This was primarily targeted at undergraduate (RIBA Part 1) but also at postgraduate (RIBA Part 2) students in order to understand industry claims that newly qualified and currently practising architects have a 'less than acceptable' attitude and approach to the subject. In order to explore this accusation, in conjunction with the RIBA and academia, a symposium was arranged at Sheffield University (25 March 2014) to bring together a series of papers from the RIBA profession, including from the RIBA Head of Education, leading academics from schools of

architecture and leading architectural practices who have all demonstrated excellence in their integration of health and safety into architecture. This was all conducted in the presence of and with contribution from the Health and Safety Executive and other interested industry representatives and academics.

Schools of architecture

The early presentations concentrated on how schools of architecture are currently trying to embed a subject that is widely interpreted by society, students and academics as 'dry, bureaucratic and legislative' into a positive, creative and wellbeing-enhancing academic culture. It was very clear that any direct reference to European and UK regulations and principles was not an appropriate vehicle to help the process; in fact quite the opposite is the case. While traditional methods of lectures, seminars and fictitious desktop projects are all commonly used, it has become increasingly apparent that these cannot clearly contextualise the culturally and ethically complex Healthy Design, Creative Safety issues into the minds of students with otherwise pass/fail course criteria to address.

Architectural practices

In later presentations the need for a greater integration with the wider construction and design practising industry was voiced by a growing body of academics and professionals, not just in the UK but internationally. Links with major clients and contractors and architectural practices embracing a variety of activities from site visits to constructionarium scenarios, inter-disciplinary groups, role play and site activity awareness seminars with time-lapse project presentations were among the proposed curriculum activities worthy of presentation to other schools.

Live projects

However, the most widely held view by the academics present was that the vehicle of 'live projects' was perceived as the most beneficial and useful teaching aid for this subject. The contextualisation of the designing and making of an actual project with a real client, design team, real site, materials and construction operatives (usually themselves) is compelling. Assimilating challenges and negotiating potential H&S barriers to overcome within a given timescale and to meet academic criteria is an ideal analogy of proportionate and practicable architecture embedding Healthy Design, Creative Safety. Learning by doing captures a life-changing experience of team interaction and actual, rather than fictitious, project delivery. While not necessarily including all regulatory H&S issues (e.g. the Principles of Prevention) this gives a realistic context to observe, discuss and negotiate all relevant influencing factors, including the legislative CDM frameworks actually underpinning these analogous activities for future real-life projects.

Overview and future implementation
The original RR925 report highlighted a number of recommendations for pursuit in future implementation phases of the CIC project, and the symposium consolidated this need together with adding other important issues such as visual, interactive and teamwork approaches. This report captures all these recommendations and provides a platform for future Healthy Design, Creative Safety integration into architectural practice and academia in a professionally acceptable and industry-wide manner.

The RIBA, schools of architecture, CIC and HSE need to consider a method of delivery of the recommendations of these reports in parallel with the introduction of CDM 2015.[45]

Appendix III
Pre-2007 research by the HSE into construction accidents

The HSE's research into the causes of accidents in 2003

The HSE commissioned a number of research reports in the years running up to CDM 2007 that tried to pin down why there was such a large number of accidents in the industry, and how designers could improve their knowledge and skills to support reduction.

The key findings in RR156 *Causal factors in construction accidents* are listed below with an associated commentary:

- *Clients and designers give insufficient consideration to health and safety, despite their obligations under the CDM Regulations.*
- *Key documentation, such as the health and safety plan, method statements and risk assessments are treated as a paper exercise, having little practical benefit.* The illusion of safety.

- *Frequent revision of work schedules leads to problems with project management and undesirable time pressure.* Stress in the design workplace can manifest itself in lack of safety in design.
- *A long hours culture in the industry results in fatigue, compromised decision-making, productivity and safety.* This is common in design offices particularly.
- *Training is seen as a solution to all problems, but with content often superficial.* Targeted training is required.
- *There have been improvements in safety culture over recent years, but safety still has to compete with other priorities.* Yes, this is the nature of design and construction projects.

The research found that:

— *Poor communication within work teams contributed to some accidents, due to the physical distance between work colleagues or high levels of background noise.* This can easily apply to design teams too, both practically and metaphorically.

— *Designers, suppliers and purchasers of equipment appear to give insufficient attention to the safety of users.* This was not intended to describe architectural designers, but they are included in the designer category, perhaps distorting the figures.

— *It was judged that up to half of the 100 accidents could have been mitigated through a design change and it was found that, despite CDM, many designers are still failing to address the safety implications of their designs and specifications.* The accusational stereotyping of this statement needs more analysis and consideration. A post-event evaluation can easily identify design changes that could lead to improved safety, but what was the agreed tolerable level of safety during the design stage? Who had ownership of those design decisions? The aspiration to design out all risk is impossible, and the tolerable risk that remains, SFARP, must be managed by the contractor team.

— *Many of the incidents were caused by commonplace hazards and activities that will continue to occur on site whatever design changes might be made. The widespread presence of the many generic safety risks accompanying construction needs to be tackled before the benefits of design improvements will be realised.* These trade-related, generic or 'trivial risks' need to be addressed by the trades and contractor groups.

— *Together, these factors point to failings in education, training and safety culture in the industry. A large majority of those working in construction, both on and off site, continue to have only a superficial appreciation of health and safety considerations.*

The report recommended that:

— *Responsibility for safety needs to be owned and integrated across the project team, from designers and engineers through to skilled trade personnel and operatives.* This is the intention of better collaborative working.

— *Where safety depends on communication and coordination, it is important that a robust safe system of work is established.* This underpins the need for better communication methods.

— *There is a need across the industry for proper engagement with risk assessment and risk management. Emphasis should be on actively assessing and controlling risk, rather than treating risk assessment as merely a paper exercise.* This is fundamental to the purpose of this book.

— *Greater opportunity should be taken to learn from failures, with implementation of accident investigation procedures, both by employers and HSE, structured to reveal contributing factors earlier in the causal chain.* Case studies to be collated by designers and hosted by a professional institute such as the RIBA.

— *It is important that 'safety' is disassociated from 'bureaucracy'.* New methods of assessment are required?

— *Frequently, safety does not have to come at a price. Where there are cost implications, however, regulatory bodies and trade associations* (and professions) *should work to make sure there is a level playing field.*[46]

Most of these recommended changes depend on achieving widespread improvement in understanding of health and safety by all contributors. Risk education is needed as well as training, to promote intelligent knowledge rather than unthinking rule-based attention to safety.

The need for health and safety information for designers in the construction industry, 2003

A study was carried out for the HSE to establish what designers' current behaviour was regarding health and safety issues in their designs, and their attitudes towards the CDM Regulations. This was done to establish what information they needed. Some quotations are listed below.

— *The CDM Regulations have given designers a responsibility which has made them think more about health and safety, for example, avoiding toxic floor glue in confined spaces.*

— *Designers now ensure that their 'backs are covered' regarding CDM.* This demonstrates a very risk-averse attitude to safety and encourages a 'malicious or mischievous compliance' approach of paperwork for its own sake just to show that something was done.

— *The health and safety plan picks up obvious risks.* Yes, the construction phase H&S plan should identify these issues, but the pre-construction information (PCI) should identify the significant ones.

— *Some designers provide operational and maintenance manuals which contain particular details such as maintenance areas and roof access, but unfortunately they are not* comprehensive. *For example, they may not include window cleaning.* This is variable across the industry but greatly improved in recent years.

— *The general opinion was that if it's not compulsory to add health and safety features, why do it?* The need must be clarified to all types of designers.

— *Adding health and safety items, such as anchor points, costs money and takes up time on the project, so these are the first things to be cut back.* Value engineering with safety needs to be integrated into all projects SFARP.

— *Time is spent complying with British Standards and Building Regulations, so spending more time on additional health and safety matters is not considered worthwhile.* A simpler set of guidance to CDM is required.

— *CDM Regulations are seen as a burden: 'more forms to fill in'.* The non- or less-bureaucratic approach was tried in CDM 2007 perhaps unsuccessfully, so now the industry should embrace it with CDM 2015.

— *Many designers were not aware of some of the very helpful published information on health and safety. For example, one designer did not know that the Health and Safety Executive has a website.* Greater access to comprehensible web information is required.

— *It was felt that it is difficult to find out about, and obtain, health and safety information.* The RIBA and other institutions need to address this issue for all sizes of practices due to cost and mass of information.

— *Also, some information within the publications is not succinct, so it takes too much time to find the facts required.*[47]

As a result of this designer-based research the HSE commissioned a peer review of designers. Unfortunately this review failed to recognise that there is a vast difference in approach to design between structural and civil engineers and architects, as well as other designers, who have much broader considerations, including visual appearance.

A peer review report on the causes of construction accidents, 2004
The HSE commissioned a task group of designers to research the reason for the industry's high accident rates, particularly targeting the role designers can play

in assisting their reduction. The conclusion is listed below with a commentary.

> The Health and Safety Executive is committed to making a fundamental reduction in the number of deaths, injuries and cases of ill health in construction. There is a view held by some of the industry and underpinned by Regulations that designers could make a significant difference. The key changes required are for designers to design structures that are safer and healthier to build, maintain and demolish. Clearly operational issues must be considered as well since they have a major effect on maintenance capability.
>
> The aesthetic, cultural and societal requirements of architects need to be considered in the safety considerations and decision-making framework, which was forgotten here.
>
> There are many in the industry, and in particular in the design community, who remain unconvinced by the arguments that designers can and should make a difference to the way they work. The purpose of this research package was to analyse actual incidents with respect to designer involvement.
>
> It is clear that designers can make a difference, but the effort needs to be proportionate SFARP.
>
> The author has chosen to track personal views of the research for the reader as this was judged helpful. In particular a certain amount of cynicism towards the arguments for real intervention by designers was in place at the beginning of the programme. Long before the end the author became completely convinced of the enormous importance of the need for radical change amongst the design community.

Most of the design community are happy to engage and agree proportionate responses with other duty holders in the context of all the other factors, but not a disproportionate over-bureaucratic approach.

> The Report concludes that almost half of all accidents in construction could have been prevented by designer intervention and that at least 1 in 6 of all incidents are at least partially the responsibility of the lead designer in that opportunities to prevent incidents were not taken.[48]

This accusation is not fully substantiated by the evidence, particularly in view of the designers of materials and equipment included. Given that design underpins most issues in construction to one degree or another it is surprising that the conclusion was not 100% involvement, but the key is recognising what is 'significant' designer involvement and what are normal construction safety issues.

Accident rates

The United Kingdom construction industry has one of the lowest accident rates in the world following generally declining rates over recent decades. Latterly, however, a levelling off has been observed and there remain various categories of seemingly intractable accidents.

CDM 1994

For some years there has been a belief that early contributions to the construction and building processes from both clients and designers could make a radical improvement to the construction processes during the whole life of a structure. Anecdotal evidence from industry showed that the construction and building industry is capable of delivering safe construction but that it regularly fails to do so. CDM put new duties on clients and designers and introduced a new statutory appointment of Planning Supervisor. The concept behind CDM was one of teams of competent appointees providing appropriate information throughout the life of the project for use by those who had the capacity to influence health and safety for good or ill. There was also a requirement to allow for adequate resources in all senses to achieve the same ends.

The opportunities presented by CDM would seem to be clearly apparent, based as they are on sound project management philosophy and holistic risk management.

The regulations were, however, generally considered by consultants and advisors in their narrowest sense and frequently not read or applied in conjunction with the Construction HS & Welfare Regulations or other relevant regulations, without which their application becomes meaningless.

Further, the Regulations were not so ordered as to make duty holders' duties easily apparent to the vast numbers of those who were obliged to wrestle with legal terminology for the first time.

Difficulties with Designers Duties – 1994 & 2007

The requirements of CDM Designers Duties have not been effectively managed by some parts of industry. Various reasons for this may exist.

— The wording of the regulation is insufficiently precise to set standards in relation to legal duties.

— There has been an assumption that CDM could stand alone without an understanding of building, construction and maintenance processes, including demolition, and of other requirements such as operational constraints. These other factors are often overlooked to the detriment of decision making.

— Many designers are either unaware of, or not up to date in, modern construction and building processes. For them to make any real contribution to safety and health they clearly need to understand where the challenges are that face those who will construct. This is an area that has been recognised by the RIBA for greater education, but proportionately.

— There has been an assumption that the regulation demanded risk assessment now commonly referred to as DRA or Design Risk Assessment. Generally the teaching of CDM to the industry has been conducted by health and safety professionals with experience in contractor risk assessments. They have tended to translate this across to the design community. In fact the Regulation makes no reference to risk assessment nor is the duty best approached by the same methods as contractor risk assessments, being rather a design process. Most DRAs are poorly conducted, retro-fitted, contractor risk assessments. The industry needs to accept that design risk assessments as previously attempted are not an appropriate method of undertaking analysis of relevant risks and the proposed integration of safety measures in design as a result.

— Many of the procurement routes, particularly those facing architects, make early intervention difficult from a commercial perspective.

— The fear of criminal action has resulted in production of excessive paperwork as an attempt to manage liability. In fact such paper trails are generally of poor quality and do little other than add to costs. They do not reduce liability unless they are effective.

Difficulties for the HSE

— HSE field inspectors are experts in the law of health and safety and its enforcement. Design is, however, a complex professional discipline requiring years of training and experience. For inspectors to challenge decisions taken by designers or to ask why alternatives have not been considered is not possible except for those inspectors with a specialist background in the appropriate discipline. Even within the industry there is a considerable range of specialist disciplines at work and the provision of competent inspectors to match every such situation is not tenable to industry.

Industrywide initiatives

The Deputy Prime Minister, John Prescott, held a construction health and safety summit where *he challenged industry to make commitment to improvement.*

Rethinking Construction and its daughter report, Rethinking Health and Safety in Construction were produced.

Designers were challenged to make a more positive contribution to health and safety in construction.[48]

As a result of recent work between the RIBA, CIC and the HSE along with architectural educators, great steps are being taken to understand the issues raised in this section. While it is accepted by the leaders of the design fraternity, it is difficult to embed this approach throughout the entire architectural profession without considerable time and effort and to engage the diverse types of design professionals. A national symposium and a series of lectures to all the UK RIBA regions held in 2014, together with this guide, is hopefully a significant step towards the embedding of a new enlightened approach to CDM.

References and bibliography

1. Directive 92/57/EEC – Temporary or mobile construction sites (24 June 1992), p. 2

2. Health and Safety Executive Construction Industry Advisory Committee (CONIAC) Paper Number M1/2009/3, 'CDM 2007 Evaluation Issues', a paper by the CONIAC Secretariat, cleared by Philip White, Chief Inspector of Construction, on 9 March 2009. Used in accordance with the 'Open Government Licence', HSE

3. *The Approved Code of Practice (ACOP): Construction (Design and Management) Regulations 2007*, HSE (London, 2007), Introduction

4. Judith Hackitt, CBE, HSE Chair, quote taken from NEBOSH graduation speech, 28 June 2010. Used in accordance with the 'Open Government Licence', HSE

5. A list of key objectives from *The Approved Code of Practice (ACOP): Construction (Design and Management) Regulations 2007*, HSE (London, 2007)

6. Professor Ragnar E. Löfstedt, 'Reclaiming health and safety for all: An independent review of health and safety legislation', November 2011

7. CIC/ICE, Löfstedt Review of Health and Safety: CIC/ICE Response, 2011, http://cic.org.uk/admin/resources/1379952549-cicice-response-to-the-lofstedt-reviews-call-for-evidence-final.pdf, accessed 05/03/2015

8. *Legal (L) Series guidance on the Construction (Design and Management) Regulations 2015*, HSE (London, 2015)

9. Dr Timothy Walker, Director General HSE, Lecture 'Reducing Risks, Protecting People – Decision-making on the Basis of Risk', 28th January 2003, held at UCL, Copyright University of Bath, http://www.bath.ac.uk/management/cri/pubpdf/Occasional_Lectures/8_Walker.pdf, accessed 05/03/2015

10. Hierarchy of Hazard Control, HSE Leadership and worker involvement toolkit, 'Management of risk when planning work: The right priorities'. Developed by the construction industry's Leadership and Worker Engagement Forum. Hosted by HSE, November 2011

11. The Construction Industry Training Board, *The Construction (Design and Management) Regulations*

2007: Industry Guidance for Designers 2007, CDM07/4 (ConstructionSkills, 2007), ISBN: 978-1-85751-236-6, pp. 21 and 22

12. *Industrial strategy: government and industry in partnership*, Department for Business, Innovation and Skills (London, 2013)

13. Reprinted by permission of the Publishers from the Introduction to *Safety-I and Safety-II* by Erik Hollnagel (Farnham: Ashgate, 2014). Copyright © 2014

14. HSG144, *The safe use of vehicles on construction sites: A guide for clients, designers, contractors, managers and workers involved with construction transport*, HSE (London, 2009)

15. Building Regulations, http://www.planningportal.gov.uk/buildingregulations/

16. Building Regulations Approved Documents, http://www.planningportal.gov.uk/buildingregulations/approveddocuments/

17. *Preventing catastrophic risks in construction*, Research Report 834, prepared by CIRIA and Loughborough University for the Health and Safety Executive, HSE (London, 2001)

18. A. Gilbertson, J. Kappia, L. Bosher and A. Gibb, CIRIA Publication C699: *Guidance on catastrophic events in construction*, CIRIA (London, 2011)

19. BS 5975:2008 Code of practice for temporary works procedures and the permissible stress design of falsework, BSI (London, 2008)

20. HSG168 Fire safety in construction, HSE (London, 2010)

21. UKTFA *Design guide to separating distances during construction, Part 1 to 3 (For timber frame buildings, above 600m^2 total floor area)*, Version 2.1 – December 2012. See also UKTFA *16 Steps to Fire Safety: Promoting best practice on Timber Frame construction sites*, July 2008. http://www.structuraltimber.co.uk/information-centre/technical-library/site-safe/

22. HSE Press Release E:01504, 30th January 2004, http://www.hse.gov.uk/press/2004/e04015.htm

23. L143: *Managing and working with asbestos*, HSE (London, 2012)

24. Control of Asbestos Regulations 2012, http://www.legislation.gov.uk/uksi/2012/632/contents/made

25. HSG150: Health and safety in construction, HSE (London, 2006)

26. BREEAM, http://www.breeam.org/

27. BSI PAS 128 Specification for underground utility detection, verification and location, BSI (London, 2014)

28. L101 Safe work in confined spaces, HSE (London, 2013)

29. Confined Spaces Regulations 1997, http://www.legislation.gov.uk/uksi/1997/1713/contents/made

30. BS 8560:2012 Code of practice for the design of buildings incorporating safe work at height, BSI (London, 2012). Permission to reproduce extracts from British Standards is granted by BSI Standards Limited (BSI). No other use of this material is permitted. British Standards can be obtained in PDF or hard copy formats from the BSI online shop: http://www.bsigroup.com/Shop

31. BS 7036 Code of practice for safety at powered doors for pedestrian use, BSI (London, 1996)

32. Gate Safe, http://gate-safe.org

33. Building Regulations Approved Documents K and N, http://www.planningportal.gov.uk/buildingregulations/approveddocuments/

34. The Construction Industry Training Board, *Industry Guidance for Principal Designers*, 2015, Annex E, CDM Red, amber, green (RAG) lists, RAG

35. Personal protective equipment (PPE), http://www.hse.gov.uk/toolbox/ppe.htm

36. Judith Hackitt, CBE, HSE Chair, quote taken from 'Engineering safe sustainable solutions – from cradle to grave', a speech given at the Institute of Mechanical Engineers, June 2012. Used in accordance with the 'Open Government Licence', HSE

37. Peter Stacey, Andrew Thorpe and Paul Roberts, *Levels of respirable dust and respirable crystalline silica at construction sites*, Research Report 878, HSE (London, 2011)

38. Notes made by the author during the 28th meeting of the CIC Health and Safety Committee, held on 20th November 2014 at the CIC Boardroom, Building Centre, 26 Store Street, London, WC1E 7BT

39. BSI PAS 91 Construction prequalification questionnaires, BSI (London, 2013)

40. Construction Skills Certification Scheme, http://www.cscs.uk.com

41. Construction Industry Advisory Committee (CONIAC), http://www.hse.gov.uk/aboutus/meetings/iacs/coniac/

42. Construction Industry Council (CIC), http://www.cic.org.uk

43. Designers' Initiative on Health and Safety (DIOHAS), http://www.diohas.org.uk

44. Tony Putsman and Paul McArthur, *Practical Guide to Using the CDM Regulations 2015*, ICE Publishing (London, 2015)

45. Leo Care, Daniel Jary and Dr Rosie Parnell, *Healthy Design, Creative Safety: Approaches to health and safety teaching and learning in undergraduate schools of architecture*, Research Report 925, HSE (London, 2012)

46. *Causal factors in construction accidents*, Research Report 156, prepared by Loughborough University and UMIST for the Health and Safety Executive, HSE (London, 2003)

47. Miscellaneous extracts from Janet Davison, *The development of a knowledge based system to deliver health and safety information to designers in the construction industry*, Research Report 173, HSE (London, 2003)

48. Liz Bennett BSc PGCE CEng MICE MIOSH FRSA, *Peer review of analysis of specialist group reports on causes of construction accidents*, Research Report 218, HSE (London, 2004). Used in accordance with the 'Open Government Licence', HSE

Index

Note: page numbers in italics refer to figures

abrasions and cuts 111
accident causes 152–4, 155–9
ACOP (Approved Code of Practice) 37, 44
ALARP suite of guidance 24–5
alterations to buildings 119
animals 105
Approved Code of Practice (ACOP) 37, 44
architectural practices 150
architectural profession 5–6
architectural schools 150
'as low as reasonably practicable' (ALARP) 24–5

biological hazards 105
building alterations 119
Building Regulations 100
building users and visitors safety 115–17
buried objects 107
burns hazard 117

carcinogenic diseases 111
card schemes for competency 146
CDM coordinators (CDM 2007) 37, 38, 50–3

CDM Options Analysis Matrix 80, *87*, *88*, *89*, *92*, *93*, *96*
CDM Regulations
 changes in CDM 2007 17–18
 comparison between CDM 2007 and 2015 35–73
 flowchart *43*
 General Principles of Prevention 26–9
 key elements 36
 reviews of previous Regulations 12–15, 19–25
CDM Visually 77–80
CIC (Construction Industry Council) 21–2, 147, 149
cladding 112–13
client duties 39, 45–9
collision hazard 116
competence 37, 44, 62, 70–3, 145–6
component handling 110
component hazards 111
confined spaces 108–9
CONIAC (Construction Industry Advisory Committee) 12, 147
Construction Dust Partnership 138, 148
Construction Industry Advisory Committee (CONIAC) 12, 147
Construction Industry Council (CIC) 21–2, 147, 149

construction loadings 110, 113
construction phase plan 60
construction phasing *95*, 105
Construction Skills Certification Scheme (CSCS) 146
contamination 107
contractors' duties 41, 64–7
cooperation 58, 62
coordination 11, 51, 53, 58
crushing hazard 115
CSCS (Construction Skills Certification Scheme) 146
cuts and abrasions 111
cutting masonry 132, *137*, 138, *140*

deleterious materials 111
demolition 18, 104, 119, 122
design concept visualisation 79, *81*, *83*, *86*
design coordination 11, 51, 53, 58
design education 149–51
design hazard analysis 78, *82*, *84–5*
design hazard awareness and risk identification checklists 101–19
design hazard logs 128–9
design hazard management register 126
design innovation 77, 132
design integration with health and safety 5–6

design reviews 58, *90–1*, 130–2
design risks 32, 79, 99 (*see also under* risk)
designers 21–2, 40, 54–9
Designers' Initiative on Health and Safety (DIOHAS) 148
dismantlement 119
documentation 57–8, 100
domestic clients 37, 39, 47–8
drowning hazard 109, 115, 117
dust 104, 114, 138
duty holders 39–69

education 149–51
electric shock hazard 106, 118
electrical interference 105
Eliminate Reduce Inform – Control 32, 52, 56–7
emergency evacuation 108
equipment risks 114
erecting structures 109–10
ERIC (Eliminate, Reduce, Inform and Control) 32, 52, 56–7
ethical approach 4
EU Directive 11
excavation 107
explosion hazards 107, 109
eye injuries 111

façade access 112, 116, *136*
falling objects/materials 110, 113
falls 117–18
falls from height 108, 110, 112, 116, 118, 133–4

falls on slopes 106
falls through fragile materials 113, 118
finishes risks 114
fire hazards 109, 113, 114, 115
fire protection 102, 106, 108, 109
first-aid 62
flat roof maintenance access *135*
fragile materials 113, 118
fumes 109, 114
furnishings risks 114

gas hazards 109
General Principles of Prevention 26–9
glazing 112–13, 116, 122

hazard awareness and risk identification checklists 101–19
hazard awareness and risk management matrix *140*
hazard awareness and risk management register 120
hazard identification tools 100–29
hazardous materials/ substances 114, 119
Health and Safety at Work Act 1974 14, 18, 22, 99
Health and Safety Executive (HSE) 10, 146–7
ALARP suite of guidance 24–5
Reducing Risks, Protecting People' (R2P2) 23–4
research into the causes of accidents 152–4, 155–9

research reports on CDM 1994 12
reviews of CDM 2007 14–15, 154–9
health and safety file 47, 49, 63
hierarchy of hazard control 30–2

information overload 100
information provision 60, 61–2, 65
innovation 77, 132
Institute of Civil Engineers (ICE) 21–2
in-use risks 115–17

lifting *see* manual handling
loadings 110, 113
Löfstedt report 19–20

maintenance of building 57, 118–19, *135* (*see also* façade access; roof access)
Management of Health and Safety at Work Regulations 1999 18
manual handling 108, 110, 111, 119, *139*
masonry cutting 132, *137*, 138, *140*
materials hazards 104, 111–12 (*see also* hazardous materials/ substances)
minimising risk 132–4
moral approach 4
musculoskeletal injuries 107, 108, 110, 111, 114, 116, 119, *139*

noise 103, 114

operatives *see* workers

PAS 91: 2013 145–6

pedestrian risks 106
personal safety 117
planning for health and safety 60, 61
planning supervisor (CDM 1994) 50–3
plant and equipment 62, 66, 114, 119
plasterboard lifting *139*
post-construction stage 49
pre-qualification questionnaires (PQQ) 70–1, 145–6
principal contractors 40–1, 60–3
principal designer (PD) 37–8, 40, 50–3
professional competence *see* competence
project phasing *94*, 105
proportionate approach 10, 14, 20, 57
Publicly Available Specification 91 (PAS 91) 145–6

'reasonably practicable' *see* 'so far as is reasonably practicable' (SFARP)
red, amber and green lists 121–4
Reducing Risks, Protecting People (R2P2) 23–4
refurbishment 102, 119
respiratory injuries 104, 111
RIBA (Royal Institute of British Architects) 149
risk assessments 19, 22, 99
risk identification tools 100–29
risk minimisation 132–4
risk registers 120, 126

roof access 27, *81*, 112, *135*
rope access *136*

Safety Schemes in Procurement (SSIP) 146
Safety Visually 77–80
Safety-I and Safety-II 76–7
scald hazard 117
securing the site 60, 64, 104, 105
services 106–7
SFARP ('so far as is reasonably practicable') 19, 20, 22–3
site access 104
site induction 60, 65
site management 60–1, 65–6
site security 60, 64, 104, 105
site set-up 29, *93*, *94*, 104
skin diseases 112
slip hazard 117
'so far as is reasonably practicable' (SFARP) 19, 20, 22–3
SSIP (Safety Schemes in Procurement) 146
stomach infection 106
structural collapse 102, 110, 112, 119
subcontractors 64
supervision 61, 65

target-setting legislation 99–100
temporary instability 110, 112
temporary support 110

threshold for appointment of coordinators 37
'tolerability of risk' (TOR) 23, 24
training 64, 70–1, 149–51
trapping hazard 116
trip hazards 117

uncontrolled collapse 119
utilities 106–7

vegetation 105
vibration 103, 114

water hazards 103, 115, 117
weather 103
welfare facilities 47, 48, 52, 60, 62, 104
wind hazards 103
window cleaning access 112, 122, *136*
Work at Height Regulations 2005 20
workers 41
    information provision 61–2, 65
    management of 61–2, 65–6, 68–9
    training 64, 70–1
working from height 112, 116, 118, 132, 133
working over water 103, 115, 117
working platforms 27, 108
working spaces 28, 108

Zero-HARM hazard awareness and risk management register 120

Image credits

Jim Arbogast (PhotoDisc V131: Workplace)	34
Adam Ciesielski	140 (X-ray image)
Jack Hollingsworth (PhotoDisc V131: Workplace)	142
Iddon, J. and Carpenter, J. Safe access for maintenance and repair. Guidance for designers second edition 2009, C686, CIRIA, London (ISBN: 978-0-86017-686-2). Go to: www.ciria.org	135
Interserve Construction Limited	94
N Jones/Probst Handling Equipment	140 (construction images)
Robert Linder	140 (paving images)
Javier Pierini (PhotoDisc V131: Workplace)	8
Photomondo (PhotoDisc V131: Workplace)	2, 16, 74, 98
Scott Brownrigg	Front cover, 24, 31, 43, 78, 79, 81–85, 90–93, 95–96, 136 (top left, bottom), 137, 139
Southampton Solent University SSU New Teaching Building images reproduced with kind permission	86–89
Total Access (UK) Limited Photograph provided with kind permission	136 (top right)